W0232152

Lecture Notes in Control and Information Sciences

Edited by A. V. Balakrishnan and M. Thoma

Vol. 1: Distributed Parameter Systems: Modelling and Identification
Proceedings of the IFIP Working Conference, Rome, Italy, June 21–26, 1976
Edited by A. Ruberti
V, 458 pages. 1978

Vol. 2: New Trends in Systems Analysis
International Symposium, Versailles, December 13–17, 1976
Edited by A. Bensoussan and J. L. Lions
VII, 759 pages. 1977

Vol. 3: Differential Games and Applications
Proceedings of a Workshop, Enschede, Netherlands, March 16–25, 1977
Edited by P. Hagedorn, H. W. Knobloch, and G. J. Olsder
XII, 236 pages. 1977

Vol. 4: M. A. Crane, A. J. Lemoine
An Introduction to the Regenerative Method for Simulation Analysis
VII, 111 pages. 1977

Vol. 5: David J. Clements, Brian D. O. Anderson
Singular Optimal Control: The Linear Quadratic Problem
V, 93 pages. 1978

Vol. 6: Optimization Techniques
Proceedings of the 8th IFIP Conference on Optimization Techniques, Würzburg, September 5–9, 1977
Part 1
Edited by J. Stoer
XIII, 528 pages. 1978

Vol. 7: Optimization Techniques
Proceedings of the 8th IFIP Conference on Optimization Techniques, Würzburg, September 5–9, 1977
Part 2
Edited by J. Stoer
XIII, 512 pages. 1978

Vol. 8: R. F. Curtain, A. J. Pritchard
Infinite Dimensional Linear Systems Theory
VII, 298 pages. 1978

Vol. 9: Y. M. El-Fattah, C. Foulard
Learning Systems:
Decision, Simulation, and Control
VII, 119 pages. 1978

Vol. 10: J. M. Maciejowski
The Modelling of Systems with Small Observation Sets
VII, 241 pages. 1978

Vol. 11: Y. Sewaragi, T. Soeda, S. Omatu
Modelling, Estimation, and Their Applications for Distributed Parameter Systems
VI, 269 pages. 1978

Vol. 12: I. Postlethwaite, A. G. J. McFarlane
A Complex Variable Approach to the Analysis of Linear Multivariable Feedback Systems
IV, 177 pages. 1979

Vol. 13: E. D. Sontag
Polynomial Response Maps
VIII, 168 pages. 1979

Vol. 14: International Symposium on Systems Optimization and Analysis
Rocquentcourt, December 11–13, 1978;
IRIA LABORIA
Edited by A. Bensoussan and J. Lions
VIII, 332 pages. 1979

Vol. 15: Semi-Infinite Programming
Proceedings of a Workshop, Bad Honnef, August 30 – September 1, 1978
V, 180 pages. 1979

Vol. 16: Stochastic Control Theory and Stochastic Differential Systems
Proceedings of a Workshop of the „Sonderforschungsbereich 72 der Deutschen Forschungsgemeinschaft an der Universität Bonn"
which took place in January 1979 at Bad Honnef
VIII, 615 pages. 1979

Vol. 17: O. I. Franksen, P. Falster, F. J. Evans
Qualitative Aspects of Large Scale Systems
Developing Design Rules Using APL
XII, 119 pages. 1979

Vol. 18: Modelling and Optimization of Complex Systems
Proceedings of the IFIP-TC 7 Working Conference
Novosibirsk, USSR, 3–9 July, 1978
Edited by G. I. Marchuk
VI, 293 pages. 1979

Vol. 19: Global and Large Scale System Models
Proceedings of the Center for Advanced Studies (CAS)
International Summer Seminar
Dubrovnik, Yugoslavia, August 21–26, 1978
Edited by B. Lazarević
VIII, 232 pages. 1979

Vol. 20: B. Egardt
Stability of Adaptive Controllers
V, 158 pages. 1979

Vol. 21: Martin B. Zarrop
Optimal Experiment Design for Dynamic System Identification
X, 197 pages. 1979

For further listing of published volumes please turn over to inside of back cover.

Lecture Notes in
Control and
Information Sciences
Edited by A.V. Balakrishnan and M. Thoma

52

A. L. Dontchev

Perturbations, Approximations and Sensitivity Analysis of Optimal Control Systems

Springer-Verlag Berlin Heidelberg GmbH 1983

Series Editors
A. V. Balakrishnan · M. Thoma

Advisory Board
L. D. Davisson · A. G. J. MacFarlane · H. Kwakernaak
J. L. Massey · Ya. Z. Tsypkin · A. J. Viterbi

Author
A. L. Dontchev
Bulgarian Academy of Sciences
Institute of Mathematics
with Computer Center
P.O.Box 373
1090 Sofia
Bulgaria

ISBN 978-3-540-12463-4 ISBN 978-3-540-44402-2 (eBook)
DOI 10.1007/978-3-540-44402-2

Library of Congress Cataloging in Publication Data
Dontchev, A. L., 1948-
Perturbations, approximations, and sensitivity analysis of optimal control systems.
(Lecture notes in control and information sciences; 52)
Bibliography: p. Includes index.
1. Control theory. 2. Perturbation (Mathematics) 3. Approximation theory.
I. Title. II. Series. QA402.3.D598 1983 629.8'312 83-4740

This work is subject to copyright. All rights are reserved, whether the whole or part of the material is concerned, specifically those of translation, reprinting, re-use of illustrations, broadcasting, reproduction by photocopying machine or similar means, and storage in data banks. Under § 54 of the German Copyright Law where copies are made for other than private use, a fee is payable to "Verwertungsgesellschaft Wort", Munich.

© by Springer-Verlag Berlin Heidelberg 1983
Originally published by Springer-Verlag Berlin Heidelberg New York in 1983

2061/3020-543210

CONTENTS

INTRODUCTION

During the last three decades a number of papers and monographs have developed various methods to estimate the effect of perturbations on system performance.The respective branch of control and system sciences was called sensitivity analysis.

The measure of sensitivity, originally defined by Bode [5] was developed as a quantitative measure of the advantages of the feedback control.The sensitivity function took the form of a ratio of the variation of the system function (state) to the deviation of the data (parameter), namely, it was defined as a derivative of the state with respect to the parameter.To be specific, consider a control system described by the following ordinary differential equation

$$\dot{x} = f(x,t,u,p) \quad ,$$

where the state x and the control u are vectors with appropriate dimensions, t is the independent variable time, p is a single time-constant parameter representing the perturbed data.The control u is a given function of the time independent of the parameter p.Then the sensitivity function of the state is defined as

$$S(t,p) = \frac{\partial}{\partial p} x(t,p)$$

and can be determined by the linearized equation

$$\dot{S} = f_x S + f_p ,$$

where $f_x = \frac{\partial}{\partial x} f$, $f_p = \frac{\partial}{\partial p} f$, assuming that these partial derivatives exist. Expanding the state x(t,p) in a Taylor series about p we get

$$x(t,p + \Delta p) = x(t,p) + S(t,p)\Delta p + o(\Delta p) ,$$

i.e. an appropriate norm of the sensitivity function can be used as a measure of the change of the state vector due to small changes of the parameter.

In this case the control is independent of the current state, that is the system is in the so-called open-loop structure. If we take the closed-loop control law

$$u = R(x,t,p)$$

then the sensitivity function will be determined by the equation

$$\dot{S} = (f_x + f_u R_x)S + f_u R_p + f_p .$$

The main purpose of the sensitivity analysis originated by Bode was to compare the sensitivity of various control structures, choosing a control law which, along with other merits, reduces the error due to parameter uncertainty.

Besides the above basic definition other functions and measures of sensitivity were widely used in the literature for systems described by more or less complicated equations. Comprehensive surveys can be found in the recent monographs by Wierzbicki [78] and Rosenwasser and Yusupov [67]. These investigations, however, were based on the differentiability axiom, that is the changes of the system function were estimated by means of the Taylor expansion.Roughly speaking, the classical sensitivity metods are practically rules for differentiation of complicated mathematical expressions describing the system structure.

The concept of optimality in control theory has essentially enriched the content of the sensitivity analysis.A variety of new problems went far beyond the scope of the classical statements.The principle connection was recognized between the sensitivity formulations and the concepts of well-posedness, stability and approximation of optimization problems.At present the sensitivity analysis in optimization includes many different trends and treatments, however, the main purpose of this theory remains the same: to characterize the changes of the optimal solution (optimal value, optimal point) due to changes of the model.

This book is devoted to sensitivity analysis in optimal control. It presents the author's investigations of the recent years.

In the sequel we briefly sketch some of the basic sensitivity problems, outlining in the same time the material in the book.For clarity,as an illustration we take the classical problem of the calculus of variations

$$J(x(\cdot),p) = \int_{t_0}^{t_1} L(x(t),t,\dot{x}(t),p)dt \longrightarrow \inf \qquad \text{(CV)}$$

$$x(t_0) = x^0 , \qquad x(t_1) = x^1 ,$$

where $x(\cdot)$ is a smooth function on $[t_0,t_1]$, $\dot{x}(\cdot)$ is the derivative with respect to t, and p is a scalar parameter representing the perturbation.

According to the definition of Hadamard [34] , a mathematical problem is well-posed if its solution is continuous (in some sense) with respect to the data.The meaning of such a well-posedness in optimization is clear: suitable small changes of the data result in arbitrary small deviations of the optimal performance.Let $\hat{x}(\cdot,p)$ solve our problem (CV) and $\hat{J}(p) = J(\hat{x}(\cdot,p),p)$ be the optimal value.We say that the problem (CV) is well-posed in the sense of performance convergence if

$$J(p) \longrightarrow J(p_o) \quad \text{as} \quad p \longrightarrow p_o \quad ,$$

where $p = p_o$ defines the original or the limit problem while p corresponds to the perturbed problem.The problem (CV) is well-posed in the sense of solution convergence when

$$x(\cdot,p) \longrightarrow x(\cdot,p_o) \quad \text{as} \quad p \longrightarrow p_o \quad ,$$

having fixed the meaning of this convergence.

A part of this book is devoted to the well-posedness of optimal control problems with two typical changes of the data - the so-called regular and singular perturbations.In Chapter 2,Section 2.2, we consider a nonlinear control problem with state and control constraints. This problem contains a regular perturbation parameter,that is, the perturbed model and the limit one have the same structure.Chapter 3 studies a more sophisticated perturbation, namely, it is concerned with the order reduction of optimal control systems.The perturbation called singular is represented by a small parameter in the derivative of the state.

For our example (CV) the corresponding singular perturbation problem may have the form

$$J(x(\cdot),p) = \int_{t_0}^{t_1} L(x(t),t,p\dot{x}(t))dt \longrightarrow \inf$$

$$x(t_0) = x^0 \ , \quad x(t_1) = x^1 \ .$$

Setting $p\dot{x} = u$, the Euler equation will contain the parameter p in the derivative

$$L_x - p\frac{d}{dt}L_u = 0 \ .$$

If p is (mathematically) small, the substitution $p = 0$ defining the limit problem reduces this equation to algebraic one.Clearly,such a reduction may result in pathological effects, e.g. boundary layers.

After some preliminary discussion in sections 3.1-3.3, Section 3.4 considers the well-posedness of two classical optimal control problems with singular perturbations.

Clearly, the problem of well-posedness in optimization allows a very general treatment.Begining with the results of Berge [3] ,there is an extensive literature dealing with the well-posedness,commonly called also stability or qualitative stability, of various abstract optimization problems, see Bank et al.[2], Huard [40] and Zolezzi [82] . The approach presented in this book,although fairly general, is based on the concrete form of the optimal control problems considered.In this way we exhibit the difficulties and specify the mathematical technique.

Although the Hadamard well-posedness helps to answer a large variety of mathematical questions, it only gives a qualitative characterization of the optimal solution.Therefore, the second basic problem of the sensitivity analysis consists in estimating the changes of the optimal solution due to changes of the model.

A natural extension of the classical sensitivity formulation to optimal control problems is based on the differentiability axiom, that is the changes of the optimal solution are to be evaluated by the first term or the first several terms in the Taylor expansion. For our exemplary problem (CV) the sensitivity function of the optimal state may be defined as

$$\hat{S}(t) = \frac{\partial}{\partial p}\hat{x}(t,p_o) .$$

By a linearization of the Euler equation

$$L_x - \frac{d}{dt}L_{\dot{x}} = 0$$

we get the following <u>sensitivity equation</u>

$$L_{x\dot{x}}\dot{\hat{S}} + L_{xx}\hat{S} + L_{xp} - \frac{d}{dt}(L_{\dot{x}\dot{x}}\dot{\hat{S}} + L_{\dot{x}x}\hat{S} + L_{\dot{x}p}) = 0 ,$$

$$\hat{S}(t_o) = \hat{S}(t_1) = 0 .$$

For such <u>unconstrained</u> problems the differentiability of the optimal solution, that is, the existence of sensitivity function requires rather moderate conditions.Practically the same conditions are needed when additional equality constraints are given.The presence of inequality constraints, however, complicates considerably the situation.Let us take the example

$$\int_0^1 (x(t) - p)^2 dt \longrightarrow \min ,$$

$$x(t) \leqslant 0 \text{ for } t \in [0,1] .$$

The solution

$$x(t,p) = \begin{cases} p & \text{for } p \leqslant 0 , \\ 0 & \text{for } p > 0 \end{cases}$$

is obviously not differentiable at $p = 0$.This effect follows from the fact that a change of the parameter in a neighbourhood of zero may lead to a change of the binding constraints.Thus, the Taylor expansion about $p = 0$ doesn't apply.Nevertheless,as we shall see in Chapter 1, using the properties of the functional one can get an exact estimation for the change of the solution.

It is the primary objective of this work to estimate the sensitivity of optimal control problems under conditions weaker than differentiability.

Chapter 1 presents some basic sensitivity evaluations for abstract constrained optimization problems.Using the properties of the uniformly convex functionals we estimate the change of the optimal point by means of the functional values and the deviation of the constraining sets.In our further considerations we apply and develop this scheme for various perturbed optimal control problems.

Section 2.3 considers a convex optimal control problem with state and control constraints.We give conditions under which the optimal control is Lipschitz continuous in L^2 metric with respect to the perturbation parameter.On the assumption that the state constraints are vacuous we refine this estimate to L^∞ and C metrices.

It should be noted that the problem of differentiability of the optimal value with respect to a parameter,commonly called <u>differential stability</u> , is not discussed in this book.For recent results see Gollan [33] ,Levitin [48] , Lempio and Maurer [49] and Robinson [64] .

Section 3.5 of Chapter 3 presents estimations for two singularly perturbed optimal control problems.We evaluate the convergence rate of the optimal value for the classical Lagrange problem.Next we directly apply the scheme of Chapter 1 obtaining estimations for the optimal control.

Instead by a parameter, the change of the model may be represented by a special procedure, which is introduced with a view to simplifying the problem.Such a procedure is the finite-difference approximation.Let us take up again the exemplary problem (CV).Applying the simplest Euler scheme we get the following discrete problem

$$\sum_{i=0}^{i=N-1} L(x_i, t_i, (x_{i+1} - x_i)/h) \longrightarrow \inf$$

$$x_0 = x^0 \ , \ x_N = x^1 \ ,$$

where $h = 1/N$ is the step size, $t_{i+1} - t_i = h, i = 0,1,\ldots,N-1$. This approximation is defined properly if the approximate solution converges in some sense to the solution of the initial problem (CV). Although the error evaluation is in fact a problem from the numerical analysis, it can be involved in the general framework of the sensitivity theory.

In Chapter 4 we develop the approach from chapters 1 and 2 for a discrete approximation to optimal control problems with local and integral constraints.For a problem with local state and control constraints first order convergence in the L^2 metric for the optimal control is established.This estimate is refined to the C metric for a problem with local control constraints and for a problem with mixed integral constraints.

Chapter 5 is devoted to the so-called real sensitivity problem formulated by Wierzbicki [77], for the open-loop control structure.

Omitting the differentiability axiom we give estimations for the sensitivity measure for convex constrained problems with various perturbations.As an example we consider a control system described by a partial differential equation.

From the wealth of literature which has recently become available for perturbation analysis in optimal control only a selection could be quoted.Therefore, we have restricted the bibliography to those papers, to which explicite reference is made in the text.

Acknowledgement: This book would never have been written without the support and encouragement of Professor Andrzej P. Wierzbicki. To him I extend my sincerest gratitude.I also extend my thanks to Professors K.Malanowski, M.Thoma and J.M.Ermolev and to Dr.V.Veliov for their advice, encouragement and valuable suggestions.

Asen L.Dontchev

Sofia, Dec. 1982

CHAPTER 1

ESTIMATES OF THE SOLUTIONS OF ABSTRACT OPTIMIZATION PROBLEMS

1.1. Uniform and strong convexity

Let B be a Banach space with a norm $\|\cdot\|$ and U be a convex subset of B.We remind, see Vladimirov et al. [74], that the real valued functional $J(\cdot): U \to R^1$ is uniformly convex on U if there exists a non-negative function $\delta(\cdot): R^1 \to R^1$,defined for all $t \in [0, \text{diam } U]$, $\delta(0) = 0, \delta(t_o) > 0$ for some $t_o > 0$, such that

$$J(\alpha u + (1-\alpha)v) \leqslant \alpha J(u) + (1-\alpha)J(v) - \alpha(1-\alpha)\delta(\|u - v\|)$$

for all $u,v \in U$ and $\alpha \in [0,1]$ .The function $\delta(\cdot)$ is the modulus of convexity of $J(\cdot)$.If $\delta(t) > 0$ for all $t \in (0, \text{diam } U]$, the functional $J(\cdot)$ is strictly uniformly convex.

Example 1.1. For every $p \geqslant 2$ the functional $\|u\|^p$ is (strictly) uniformly convex on B with

$$0 \leqslant \delta(t) \leqslant t^p/2^{p-2} \quad \text{for } t \geqslant 0 .$$

We briefly describe some of the basic properties of the uniformly convex functionals.

Let U be a convex and closed set in B.Consider the infimum problem

$$J(u) \longrightarrow \inf , \quad u \in U .$$

If the space B is reflexive , the (strict) uniform convexity of $J(\cdot)$ implies existence of (unique) solution

$$\hat{u} = \text{argmin } J(u) , \quad u \in U .$$

For every $u \in U$ and $\alpha \in (0,1]$ we have

$$0 \leqslant J(\alpha u + (1-\alpha)\hat{u}) - J(\hat{u}) \leqslant \alpha(J(u) - J(\hat{u})) - \alpha(1-\alpha)\delta(\|u - \hat{u}\|) .$$

This inequality yields

$$\delta(\|u - \hat{u}\|) \leqslant J(u) - J(\hat{u}) \quad , \tag{1}$$

that is, if we know the properties of the function $\delta(\cdot)$ we can estimate the distance between some $u \in U$ and the optimal point $\hat{u}$ by means of the functional values. This observation is an initial point for our analysis in the following two sections.

Denote by B^* the adjoint space of B and let $\langle\cdot,\cdot\rangle$ be the duality between B and B^*. In the sequel we shall use the following assertions: Suppose that the functional $J(\cdot)$ is bounded and uniformly convex on an open and convex set U^0. Denote by $\partial J(u)$ the subdifferential of $J(u)$ at $u \in U^0$. Then

$$J(u) \geqslant J(v) + \langle c(v).u - v\rangle + \delta(\|u - v\|) \quad , \tag{2}$$

$$\langle c(u) - c(v), u - v\rangle \geqslant 2\delta(\|u - v\|) \tag{3}$$

for every $c(u) \in \partial J(u)$, $c(v) \in \partial J(v)$ and for every $u, v \in U^0$.

If the modulus of convexity has the form

$$\delta(t) = \varkappa t^2 ,$$

where $\varkappa > 0$ is a given constant, the functional is called <u>strongly convex</u> (introduced by Poljak [61]). In this case the inequality (1) implies

$$\|u - \hat{u}\| \leqslant ((J(u) - J(\hat{u}))/\varkappa)^{0.5} .$$

In order to avoid cumbersome but not essential for our purposes assumptions, in the following two sections we limit our considerations to strongly convex functionals. Actually, this supposition can be weakened if we know the continuity properties of $\delta(t)$ at $t = 0$, see Remark 1.1 in Section 1.2. On the other hand, considering strongly convex functionals we will obtain global estimates which, as we shall see later, have direct applications in optimal control.

1.2. Problems with set constraints

Consider the following two optimization problems

$$J_p(u) \longrightarrow \inf \ , \ u \in U_p, \tag{4}$$

$$J_o(u) \longrightarrow \inf \ , \ u \in U_o, \tag{5}$$

where the objective functions $J_p(\cdot)$ and $J_o(\cdot)$ are real valued functionals defined in the Banach space B and U_p and U_o are subsets of B representing the constraints.These two problems can be regarded as two models of a given original problem, where the subscript p denotes the perturbed model and o denotes the basic model.Our purpose is to estimate the deviation of the optimal point due to changes of the model.

Let us recall that the functional $J_o(\cdot)$ is strongly convex on U_o with a constant $\varkappa_o > 0$ if

$$J_o(\alpha u + (1-\alpha)v) \leqslant \alpha J_o(u) + (1-\alpha)J_o(v) - \alpha(1-\alpha)\varkappa_o \|u - v\|^2$$

for all $u,v \in U_o$ and $\alpha \in [0,1]$.

Let co U denote the convex hull of U.In the sequel we assume that:

A1. The functional $J_o(\cdot)$ is strongly convex on $co(U_p \cup U_o)$ with a constant $\varkappa_o$.The problems (4) and (5) have solutions $\hat{u}_p$ and $\hat{u}_o$ respectively.

Notice that the strong convexity implies uniqueness of $\hat{u}_o$.

Proposition 1.1. Suppose that $\hat{u}_p \in U_o$.Then for every $u_p \in U_p$ the following inequality holds

$$\|\hat{u}_p - \hat{u}_o\| \leqslant ((J_o(\hat{u}_p) - J_p(\hat{u}_p) + J_p(u_p) - J_o(\hat{u}_o))/\varkappa_o)^{0.5}. \tag{6}$$

Proof.From (1) we have

$$\varkappa_o \|\hat{u}_p - \hat{u}_o\|^2 \leqslant J_o(\hat{u}_p) - J_o(\hat{u}_o) \ . \tag{7}$$

This inequality, combined with

$$J_p(\hat{u}_p) \leqslant J_p(u_p)$$

gives us (6),Q.E.D.

Corollary 1.1. Let $J_p(\cdot) = J_o(\cdot)$, $\hat{u}_p \in U_o$ and L_p be a constant such that for some $u_p \in U_p$

$$J_o(u_p) - J_o(\hat{u}_o) \leqslant L_p \| u_p - \hat{u}_o \| .$$

Then

$$\| \hat{u}_p - \hat{u}_o \| \leqslant (L_p \| u_p - \hat{u}_o \| / \varkappa_o)^{0.5} . \qquad (8)$$

This estimate can be interpreted in the following way: the perturbed solution $\hat{u}_p$ converges to the basic one $\hat{u}_o$ with a rate proportional to the square rooth the distance between $\hat{u}_o$ and U_p.

Example 1.2. Let $B = R^2$, $J_o(u) = (u_1 - 1)^2 + u_2^2$. The set U_o is the half plane, the extreme point of U_p coincides with the optimal point $\hat{u}_p$ for all p as $p \to 0$, $U_p \subset U_o$, see Fig.1.1.

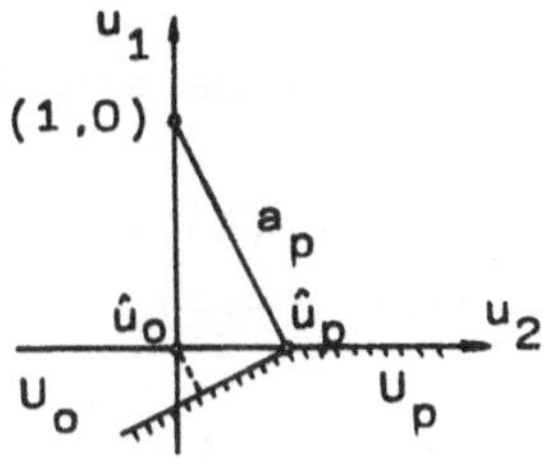

Fig.1.1.

Clearly, $\varkappa_o = 1$ and

$$\| \hat{u}_p - \hat{u}_o \|^2 = a_p \operatorname{dist}(\hat{u}_o, U_p) \quad , \; a_p \to 1 \text{ as } p \to 0.$$

Hence, the exponent 0.5 in (8) is exact.

Corollary 1.2. Let $U_p = U_o$ and $\hat{u}_p$ solve (4). Then

$$\| \hat{u}_p - \hat{u}_o \| \leqslant ((J_o(\hat{u}_p) - J_p(\hat{u}_p) + J_p(\hat{u}_o) - J_o(\hat{u}_o))/\varkappa_o)^{0.5}. \qquad (9)$$

Example 1.3. $U_p = U_o = R^1$, the scalar parameter $p \to 0$ and the functionals $J_p(\cdot)$ and $J_o(\cdot)$ are:

$$J_p(u) = 0.5u^2 + \begin{cases} 0 & \text{for } u \geqslant \sqrt{p}\,, \\ 0.5u^2 - u\sqrt{p} + 0.5p & \text{for } u < \sqrt{p}\,, \end{cases}$$

$J_o(u)$ corresponds to $p = 0$.

Observe that the functional $J_p(u)$ is Lipschitz continuous with respect to p at $p = 0$ uniformly in $u \in R^1$.Hence,the estimate (9) yields that the solution $\hat{u}_p$ is Hoelder continuous with respect to p at $p = 0$ with exponent 0.5.In our example $\hat{u}_p = 0.5\sqrt{p}$,hence the exponent in (9) cannot be improved,in general.

Now we replace A1 by the following assumption:

A2. The sets U_p and U_o are convex.The functional $J_o(\cdot)$ is convex on U_o and $J_p(\cdot)$ is strongly convex on $co(U_p \cup U_o)$ with a constant $\varkappa_p$. The functionals $J_p(\cdot)$ and $J_o(\cdot)$ are Frechet differentiable in an open set containing $co(U_p \cup U_o)$ and the derivatives $J_p'(\cdot)$ and $J_o'(\cdot)$ are continuous.There exist solutions $\hat{u}_p$ and $\hat{u}_o$.

Clearly, $\hat{u}_p$ is unique.

Let $|\cdot|$ denote the absolute value and $\|\cdot\|^*$ be the norm of the adjoint space B^*.

<u>Proposition 1.2</u>. Let $\hat{u}_o$ solve (5),Then for every $u_p \in U_p$ and $u_o \in U_o$ the following inequality holds

$$\|\hat{u}_p - \hat{u}_o\| \leqslant \|J_p'(\hat{u}_o) - J_o'(\hat{u}_o)\|^*/2\varkappa_p$$

$$+ (|\langle J_p'(\hat{u}_p), u_p - \hat{u}_o\rangle + \langle J_o'(\hat{u}_o), u_o - \hat{u}_p\rangle|/2\varkappa_p)^{0.5}. \quad (10)$$

Proof.The inequality (3) has the form

$$2\varkappa_p\|\hat{u}_p - \hat{u}_o\|^2 \leqslant \langle J_p'(\hat{u}_p) - J_p'(\hat{u}_o), \hat{u}_p - \hat{u}_o\rangle \quad . \quad (11)$$

The convexity implies that for every $u_p \in U_p$ and $u_o \in U_o$

$$\langle J_p'(\hat{u}_p), u_p - \hat{u}_p\rangle \geqslant 0 \;, \quad (12)$$

$$\langle J_o'(\hat{u}_o), u_o - \hat{u}_o\rangle \geqslant 0 \;. \quad (13)$$

Denote

$$a = \| J_p'(\hat{u}_o) - J_o'(\hat{u}_o)\|^* \ , \tag{14}$$

$$b = |\langle J_p'(\hat{u}_p), u_p - \hat{u}_o\rangle + \langle J_o'(\hat{u}_o), u_o - \hat{u}_p\rangle| \ . \tag{15}$$

Combining (11) - (13) we obtain

$$2\varkappa_p \|\hat{u}_p - \hat{u}_o\|^2 \leqslant a\|\hat{u}_p - \hat{u}_o\| + b \ . \tag{16}$$

Solving this inequality we get (10), Q.E.D.

Obviously, one can replace the Frechet derivatives by subgradients, see (3).

<u>Corollary 1.3</u>. Suppose that $U_p = U_o$. Then, if $\hat{u}_o$ solves (5)

$$\|\hat{u}_p - \hat{u}_o\| \leqslant \|J_p'(\hat{u}_o) - J_o'(\hat{u}_o)\|^*/2\varkappa_p \ . \tag{17}$$

<u>Example 1.4</u>. For the one-dimensional quadratic problem

$$J_p(u) = 0.5u^2 - pu \longrightarrow \inf$$

$$U_p = U_o = R^1 \ , \ p \in R^1 \ ,$$

we have

$$\varkappa_p = 0.5, \quad \hat{u}_p = p, \quad J_p'(\hat{u}_o) = -p, \quad J_o'(\hat{u}_o) = 0 \ ,$$

and (17) holds as equality.

<u>Corollary 1.4</u>. Suppose that $J_o(\cdot) = J_p(\cdot)$, $\hat{u}_o \in \operatorname{int} U_o$, and let L_p' be a constant such that

$$\|J_p'(\hat{u}_p) - J_p'(\hat{u}_o)\|^* \leqslant L_p'\|\hat{u}_p - \hat{u}_o\| \ .$$

Then for every $u_p \in U_p$

$$\|\hat{u}_p - \hat{u}_o\| \leqslant L_p'\|u_p - \hat{u}_o\|/2\varkappa_p \ . \tag{18}$$

Proof. From (11),(12) and the minimum condition $J_o'(\hat{u}_o) = 0$ we get

$$2\varkappa_p\|\hat{u}_p - \hat{u}_o\|^2 \leqslant \langle J_p'(\hat{u}_p), u_p - \hat{u}_o\rangle \leqslant \|J_p'(\hat{u}_p) - J_p'(\hat{u}_o)\|^*\|u_p - \hat{u}_o\| .$$

This gives us (18),Q.E.D.

Observe that the exponent in (18) is twice the exponent in (8). This results from the assumption that $\hat{u}_o \in \text{int } U_o$.

Example 1.5. Let B be a Hilbert space and $J_p(\cdot) = J_o(\cdot) = J(\cdot)$ be a quadratic form

$$J(u) = 0.5\langle u,Au\rangle + \langle b,u\rangle,$$

The operator A: $B \to B$ is linear, bounded,selfadjoint, and there exist positive constants β and L such that

$$\langle u,Au\rangle \geq \beta \|u\|^2 , \qquad \|Au\| \leq L\|u\| \qquad \text{for all } u \in B .$$

Clearly, $\varkappa_p = 0.5\beta$.Let $\hat{u}_p$ minimize $J(\cdot)$ over a convex set $U_p \subset B$ and let $\hat{u}_o$minimize $J(\cdot)$ over the entire space B.Then (18) yields

$$\|\hat{u}_p - \hat{u}_o\| \leq L\inf\|u - \hat{u}_o\|/\beta , \quad u \in U_p .$$

This estimate,known as Céa's lemma, plays an important role in the numerical analysis of variational problems,see Ciarlet [8] p.104.

Remark 1.1. It is clear that the exponent 0.5 in the above estimates results from the strong convexity assumption.It is also clear that one can replace this assumption by some weaker uniform convexity condition.For example, let us take Proposition 1.2.From (3),(12) and (13) we get

$$2\delta_p(\|\hat{u}_p - \hat{u}_o\|) \leq a\|\hat{u}_p - \hat{u}_o\| + b ,$$

where a and b are determined in (14) and (15) and $\delta_p(\cdot)$ is the modulus of convexity of $J_p(\cdot)$.The continuity properties of $\delta_p(\cdot)$ will provide an estimate for the distance between $\hat{u}_p$ and $\hat{u}_o$..

1.3. Convex programming problems

Let B, Y and Z be Banach spaces and D be a positive closed and convex cone in Z defining the relation $x \leq y$ when $x - y \in D$.In this section we specify the problems (4) and (5) from the previous section assuming that the sets U_p and U_o are determined by equality and inequality relations, that is

$$U_i = \{u \in B \ , \ F_i u = b_i \ , \ G_i(u) \leqslant 0 \} \quad , \ i \in \{0,p\} \ , \qquad (19)$$

where the operators $F_i : B \to Y$ are linear and bounded and the operators $G_i : B \to Z$ are D-convex,that is

$$G_i(\alpha x + (1-\alpha)y) \leqslant \alpha G_i(x) + (1-\alpha)G_i(y)$$

for all $x,y \in B$ and $\alpha \in [0,1]$ .We assume that the conditions A2 from the previous section hold and $G_i(\cdot)$ are Frechet differentiable on B. Then the problems

$$J_i(u) \to \inf \ , \ u \in U_i \quad , \ i \in \{0,p\}$$

are convex programming problems.

We shall denote by $\|\cdot\|$ the norms of all the considered spaces, by $|\cdot|$ the absolute value and by $\langle \cdot , \cdot \rangle$ the duality leaving to the context to fix the respective meaning.The adjoint spaces (operators) are denoted by asterisks.For simplicity and ease of notation we assume that the set of solutions of the basic problem $(i = 0)$ consists of one point $\hat{u}_0$.

Let us recall,see for instance Luenberger [50] , that the Lagrange functional for the problem considered has the form

$$L_i(u;\theta,\mu,\lambda) = \theta J_i(u) + \langle \mu, F_i u - b_i \rangle + \langle \lambda, G_i(u) \rangle \ , \qquad (20)$$

where $\theta \in R^1$, $\theta \geqslant 0$, $\mu \in Y^*$, $\lambda \in Z^*$, $\lambda \geqslant 0$.From the duality theory it follows that there exist optimal Lagrange multipliers θ^i, μ^i and θ^i such that

$$\theta^i J_i(\hat{u}_i) = \min_{u \in B} L_i(u;\theta^i,\mu^i,\lambda^i) \ , \ i \in \{0,p\} \ , \qquad (21)$$

or, in equivalent form

$$\theta^i J_i'(\hat{u}_i) + F_i^* \mu^i + G_i'(\hat{u}_i)^* \lambda^i = 0 \qquad (22)$$

and

$$\langle \lambda^i, G_i(\hat{u}_i) \rangle = 0 \ , \ i \in \{0,p\} \ . \qquad (23)$$

For notation convinience, denote

$$\varphi_i(u) = (F_i u - b_i,\ G_i(u))\ ,\ \eta^i = (\mu^i, \lambda^i)\ .$$

Proposition 1.3. The following relation holds

$$\theta^p \varkappa_p \| \hat{u}_p - \hat{u}_o \| \leqslant \| \theta^p J_p'(\hat{u}_o) - \theta^o J_o'(\hat{u}_o) + (\varphi_p'(\hat{u}_o) - \varphi_o'(\hat{u}_o))^* \eta^o \|$$
$$+ |\theta^p \varkappa_p \langle \eta^p - \eta^o, \varphi_p(\hat{u}_o) - \varphi_o(\hat{u}_o) \rangle |^{0.5}\ . \qquad (24)$$

Proof.From (21) we have

$$\theta^p J_p(\hat{u}_p) \leqslant L_p(\hat{u}_o; \theta^p, \eta^p)$$
$$\leqslant \theta^p J_p(\hat{u}_o) + \langle \eta^p, \varphi_p(\hat{u}_o) - \varphi_o(\hat{u}_o) \rangle\ . \qquad (25)$$

On the other hand, using (2) and the optimality condition (22)

$$\theta^p J_p(\hat{u}_p) \geqslant \theta^p J_p(\hat{u}_p) + \langle \eta^o, \varphi_p(\hat{u}_p) \rangle \geqslant \theta^p J_p(\hat{u}_o) + \langle \eta^o, \varphi_p(\hat{u}_o)$$
$$- \varphi_o(\hat{u}_o) \rangle + \langle \theta^p J_p'(\hat{u}_o) - \theta^o J_o'(\hat{u}_o)$$
$$+ (\varphi_p'(\hat{u}_o) - \varphi_o'(\hat{u}_o))^* \eta^o, \hat{u}_p - \hat{u}_o \rangle + \varkappa_p \theta^p \| \hat{u}_p - \hat{u}_o \|^2\ .$$

Subtracting the last inequality from (25) we obtain (24),Q.E.D.

The following example shows that the convergence rate of the perturbed solution $\hat{u}_p$ to the limit one $\hat{u}_o$ is determined mainly by the addent in the right hand side of (24).

Example 1.2.(Continuation)In the previous section we considered a problem,which can be written as

$$(u_1 - 1)^2 + u_2^2 \to \min\ ,$$
$$u_1 - u_2\sqrt{p} + p \leqslant 0\ ,$$
$$u_1 \leqslant 0\ .$$

The constraining sets for $p > 0$ and for $p = 0$ are given in Fig. 1.1.We have

$$J_o(\cdot) = J_p(\cdot),\ \varkappa_p = 1,\ \hat{u}_p = (0,\sqrt{p})\ ,\hat{u}_o = (0\,,0)\ ,$$

$$\varphi_p(\hat{u}_o) = (p\,,0)\ ,\ \varphi_o(\hat{u}_o) \ = (0\,.\,0)\ ,$$

$$L = \Theta(u_1 - 1)^2 + \theta u_2^2 + \lambda_1(u_1 - u_2\sqrt{p} + p) + \lambda_2 u_1\ ,$$

$$2\theta(u_1 - 1) + \lambda_1 + \lambda_2 \ = 0,$$

$$2\theta u_2 - \lambda_1\sqrt{p} \ = 0\ .$$

Taking $\theta^p = 0.5$ we get

$$\lambda_1^p = 1\ ,\ \lambda_2^p = 0\ .$$

For $p = 0$ both inequalities coincide, however, the number of constraints should be preserved.Setting

$$\lambda_1^o = 0,\ \lambda_2^o - 2\theta^o \ = 0$$

we obtain

$$(\varphi_p'(\hat{u}_o) - \varphi_o'(\hat{u}_o))^* \eta^o = 0\ ,$$

$$\langle \eta^p - \eta^o\ ,\varphi_p(\hat{u}_o) - \varphi_o(\hat{u}_o)\rangle = p\ .$$

The first term in (24) vanishes while the convergence rate (square rooth of p) is determined by the second term.

The efficiency of the above estimates depends on the limit properties of the Lagrange multipliers. In order to specify these properties we consider further the following sequence of infimum problems

$$J_u(u) \to \inf\ ,\ u \in U_n\ ,\ n = 0,1,\ldots,$$

where the sets U_n are defined as in (19) by F_n, b_n, $G_n(\cdot)$, and $J_n(\cdot)$ and U_n satisfy all the conditions for $J_p(\cdot)$ and U_p.

We assume in addition that:

A3. The sequence $\varkappa_n$ is bounded below by $\varkappa > 0$. The spaces Z and Y are Hilbert spaces. For every bounded sequence u_n the sequences $J_n(u_n)$, $J_n'(u_n)$ and $\varphi_n(u_n)$ are bounded. For every bounded sequence e_n, $e_n \in Y$, there exists a bounded sequence u_n such that $F_n u_n = e_n$. The cone D has nonempty interior and there exists a bounded sequence $\bar{u}_n$ and a constant $\beta > 0$ such that for every $z \in Z, |z| \leqslant \beta$, $G_n(u_n) \leqslant z$ and $F_n u_n = b_n$ for $n = 0,1,\ldots.$

<u>Lemma 1.1</u>. The sequence of solutions $\hat{u}_n$ is bounded and

$$\liminf_{n \to \infty} \Theta^n > 0 .$$

Proof. The inequality (1) yields

$$\|\hat{u}_n - \bar{u}_n\|^2 \leqslant J_n(\bar{u}_n) - J_n(\hat{u}_n) . \tag{26}$$

Furthermore

$$J_n(\bar{u}_n) - J_n(\hat{u}_n) \leqslant \|J_n'(\bar{u}_n)\| \|\bar{u}_n - \hat{u}_n\| . \tag{27}$$

Combining (26) and (27) we conclude that the sequences $\hat{u}_n$ and $J_n(\hat{u}_n)$ are bounded. Clearly, $\Theta^n > 0$. Since

$$L_n(\hat{u}_n;\Theta^n,\eta^n) \leqslant L_n(\bar{u}_n;\Theta^n,\eta^n)$$

we get

$$\langle \lambda^n, G_n(\bar{u}_n)\rangle \geqslant \Theta^n(J_n(\hat{u}_n) - J_n(\bar{u}_n)) . \tag{28}$$

Without loss of generality we suppose that

$$\Theta^n + \|\mu^n\| + \|\lambda^n\| = 1 , \tag{29}$$

for $n = 0,1,\ldots.$ Since

$$\langle \lambda^n, G_n(\bar{u}_n) - \beta\lambda^n \rangle \leqslant 0 ,$$

we have

$$\langle \lambda^n, G_n(\bar{u}_n)\rangle \leqslant -\beta|\lambda^n|^2 = -\beta(1 - \Theta^n - \|\mu^n\|)^2 \leqslant 0. \tag{30}$$

Assume that $\lim_{n\to\infty} \theta^n = 0$. From (28) and (30) we obtain

$$\lim_{n\to\infty} \|\mu^n\| = 1\,, \qquad \lim_{n\to\infty} \|\lambda^n\| = 0. \tag{31}$$

There exists a bounded sequence $\tilde{u}_n$ such that

$$F_n\tilde{u}_n = -\mu^n/\|\mu^n\| + b_n\,,$$

Hence

$$\theta^n(J_n(\hat{u}_n) - J_n(\tilde{u}_n)) \leqslant -\|\mu^n\| + \langle \lambda^n, G_n(\tilde{u}_n)\rangle\,.$$

Using (31), for sufficiently large n we come to a contradiction. This completes the proof.

In the sequel we suppose that the constraints (19) are described by a finite number of functional relations such that the derivatives associated with the binding constraints, are linearly independent. More precisely, we assume that:

A4. B is a Hilbert space, $Y = R^r$, $Z = R^m$. The cone D consists of all vectors with nonnegative components. For every bounded sequence u_n the sequence $\varphi_n'(u_n)$ is bounded and

$$\lim_{n\to\infty} \|u_n - \hat{u}_o\| = 0 \qquad \text{implies} \qquad \lim_{n\to\infty} \|\varphi_n'(u_n) - \varphi_o'(\hat{u}_o)\| = 0.$$

Moreover

$$\lim_{n\to\infty} (\|J_n'(\hat{u}_o) - J_o'(\hat{u}_o)\| + \|\varphi_n(\hat{u}_o) - \varphi_o(\hat{u}_o)\|) = 0,$$

and the matrix

$$M = [\varphi_o'(\hat{u}_o)_I \varphi_o'(\hat{u}_o)_I^*]$$

is nonsingular, where the subscript I denotes the restriction on the set of binding constraints $\varphi_o(\hat{u}_o)_i = 0$, $i \in \{1,\dots,m+r\}$.

The last requirement is known as Kuhn-Tucker regularity condition. It implies that, taking $\theta^o = 1$, the <u>normal</u> Lagrange multiplier η^o is unique.

Further, omitting the normalizing condition (29), we shall use the normal form of the Lagrange functional (20), setting $\theta^n = 1$

for n = 0,1,....Denote

$$I = \{ i \in \{1,\dots,m+r\},\ \ \varphi_o(\hat{u}_o)_i = 0 \}.$$

Then from (23) we deduce that $\lambda_i^o = 0$ for $i \notin I$.

Lemma 1.2. For n sufficiently large the normal dual multiplier η^n is unique and $\lambda_i^n = 0$ for $i \notin I$.

Proof.Applying A4 and Lemma 1.1 to the estimate (24) we immediately obtain

$$\lim_{n\to\infty} \|\hat{u}_n - \hat{u}_o\| = 0. \tag{32}$$

There exists a constant $\gamma>0$ such that for every $z = (z_i), i \in I$,

$$\gamma\|z\| \leqslant \Big\|\sum_{i\in I} \varphi_o'(\hat{u}_o)_i^* z_i\Big\|$$

$$\leqslant \Big\|\sum_{i\in I} \varphi_n'(\hat{u}_n)_i^* z_i\Big\| + \|\varphi_n'(\hat{u}_n) - \varphi_o'(\hat{u}_o)\|\,\|z\| .$$

Hence, for n sufficiently large

$$0.5\gamma\|z\| \leqslant \Big\|\sum_{i\in I} \varphi_n'(\hat{u}_n)_i^* z_i\Big\| .$$

Moreover, $G_n(\hat{u}_n)_i < 0$ for $i \notin I$.Thus, from (22),(23),(32) and the uniqueness of $\hat{u}_n$ we conclude that η^n is unique and $\lambda_i^n = 0$ for $i \notin I$, Q.E.D.

Proposition 1.4. Suppose that there exists a constant K such that

$$\|J_n'(\hat{u}_n) - J_n'(\hat{u}_o)\| \leqslant K\|\hat{u}_n - \hat{u}_o\|,$$

and

$$\|G_n'(\hat{u}_n) - G_n'(\hat{u}_o)\| \leqslant K\|\hat{u}_n - \hat{u}_o\|.$$

Then there exists a constant C such that for n sufficiently large

$$\|\hat{u}_n - \hat{u}_o\| + \|\eta^n - \eta^o\| \leqslant C(\|J_n'(\hat{u}_o) - J_o'(\hat{u}_o)\|$$

$$+ \|\varphi_n'(\hat{u}_o) - \varphi_o'(\hat{u}_o)\| + \|\varphi_n(\hat{u}_o) - \varphi_o(\hat{u}_o)\|).$$

Proof. From (22) and Lemma 1.2 we have

$$\sum_{i \in I} \varphi_n'(\hat{u}_n)_i^* \eta_i^n = -J_n'(\hat{u}_n)$$

for n = 0 and for n sufficiently large.Applying Lemma 1.1 and A4 we get

$$\|\eta^n - \eta^o\| \leq C_1(\|\hat{u}_n - \hat{u}_o\| + \|J_n'(\hat{u}_o - J_o'(\hat{u}_o)\| + \|\varphi_n'(\hat{u}_o) - \varphi_o'(\hat{u}_o)\|).$$

where C_1 is independent of n.Combining this inequality with (24) we complete the proof.

Observe that the exponent in the above estimate is twice the exponent in (24).This is a consequence of the Kuhn-Tucker regularity condition which enables us to estimate the distance between the dual variables.

In our example 1.2 we have two identical constraints,thus, the regularity condition does not hold.Clearly, this comes from the assumption that the number of constraints is not perturbed.

CHAPTER 2

REGULAR PERTURBATIONS

2.1. Introduction and notation

In this chapter we consider optimal control problems containing a perturbation parameter. The perturbation is called regular, in contrast to the singular one, since it does not provide a change of the model structure. Singularly perturbed problems will be studied in the next chapter.

The chapter is organized as follows: Section 3.2 is concerned with the performance well-posedness of constrained control problems. At first we give a short review of the mathematical technique. The scheme is rather standard; it has been used more or less explicitly in a number of works dealing with various time-optimal, linear, linear-quadratic and convex problems, e.g. Gabasov and Kirillova [24], Kirillova [43], Gičev [26], Petrov [60], Tchukanov [69] and Zolezzi [81]. Applying the concept of relaxation, in § 2.2.2 we develop this sheme for a nonconvex problem with state and control constraints.

Section 2.3 deals with the convergence rate of the perturbed optimal control to the basic one for constrained convex optimal control problems. Using Lagrange duality theory and some ideas from the previous chapter we derive estimations for the optimal solution. We distinguish four cases: § 2.3.1 - the constraints are arbitrary convex sets in the spaces of the control functions and the trajectories; § 2.3.2 - the constraints are given by inequalities; § 2.3.3 - the binding constraints satisfy an independence condition, and § 2.3.4 - there are no state constraints. At the end we consider as an example a problem with fixed initial and final state.

Throughout this and the following chapters we shall denote by $|\cdot|$ the Euclidean norm and by the superscript T the transposition. The following notation is used for spaces of functions $f(\cdot)$ defined on the interval $[0,1]$ with values in R^n :

$L^p(R^n)$ - functions with

$$\|f\|_{L^p} = \left(\int_0^1 |f(t)|^p dt\right)^{1/p} < +\infty \quad , \ 1 < p < \infty ;$$

the $L^2(R^n)$ norm of $f(\cdot)$ is denoted simply by $\|f\|$; the scalar product in $L^2(R^n)$ of the functions $f(\cdot)$ and $g(\cdot)$ is denoted by $\langle f,g\rangle$, i.e.

$$\langle f,g\rangle = \int_0^1 f(t)^T g(t)dt \ .$$

$L^\infty(R^n)$ - essentially bounded functions with the norm

$$\|f\|_{L^\infty} = \text{vraisup}\{|f(t)| \ , \ 0 \leqslant t \leqslant 1\}.$$

$BV(R^n)$ - functions of bounded variation that are left continuous on $[0,1)$; $V_a^b(f)$ denotes the variation of $f(\cdot)$ on $[a,b]$.

$A(R^n)$ - absolutely continuous functions.

$H^1(R^n)$ - absolutely continuous functions with first derivative in $L^2(R^n)$, a scalar product

$$(x,y) = \langle x,y\rangle + \langle \dot{x},\dot{y}\rangle \ ,$$

and a norm generated by this scalar product.

$C(R^n)$ - continuous functions with the usual maximum norm.

$C^p(R^n)$ - functions with continuous derivatives through order p.

2.2.Well-posedness

2.2.1. An outline of the method

The next paragraph gives conditions for performance well-posedness of nonconvex control problems.Chapter 3 contains further results concerned with the well-posedness of singularly perturbed problems. The mathematical technique that is used arises from the following general scheme.

To be concrete, consider the following problem

$$J\ (u(\cdot)) = g(x(1)) \longrightarrow \inf \tag{1}$$

subject to

$$\dot{x} = \varphi_\varepsilon(x,u,t) \;,\; x(0) = x^0 \;, \tag{2}$$

$$u(t) \in V \text{ for almost every } t \in [0,1]$$

$$x(\cdot) \in A(R^n) \;,\; u(\cdot) \in L^2(R^m) \;,$$

where the parameter ε represents the perturbation of the data.The basic problem corresponds to $\varepsilon = 0$.For simplicity, let $\varepsilon \geqslant 0$.

Denote

$$\hat{J}_\varepsilon = \inf J_\varepsilon(u(\cdot)) \;.$$

Then the problem will be well-posed in the sense of performance convergence if

$$\lim_{\varepsilon \to 0} \hat{J}_\varepsilon = \hat{J}_0 \;.$$

For clarity, we specify the assumptions throughout the presentation of the method leaving the precise formulations to the reader.

Let $\varepsilon_k \to 0$ as $k \to \infty$ and let $\tilde{u}_k(\cdot)$ be a sequence of feasible controls (if any) for which

$$J_{\varepsilon_k}(\tilde{u}_k(\cdot)) \leqslant \hat{J}_{\varepsilon_k} + \varepsilon_k \quad ,k = 1,2,...$$

Let $\tilde{x}_k(\cdot)$ correspond to $\tilde{u}_k(\cdot)$ and ε_k according to (2).Then, under some mild conditions one can deduce that the sequence of the derivatives $\dot{\tilde{x}}_k(\cdot)$ contains a L^2 - weakly convergent subsequence.This is true for example when $\varphi(\cdot)$ is continuous,the set V is compact and there exists a compact set $X \subset R^n$ containing every trajectory of (2).

Let $p(\cdot)$ be a weak limit of $\dot{\tilde{x}}_k(\cdot)$ and let

$$\tilde{x}(t) = x^0 + \int_0^t p(s)ds \;.$$

Clearly, the sequence of the trajectories $\tilde{x}_k(\cdot)$ converges uniformly to $\tilde{x}(\cdot)$.

If the limit function $\tilde{x}(\cdot)$ belongs to the set of trajectories for $\varepsilon = 0$, then there exists a control $\tilde{u}(\cdot)$,corresponding to $\tilde{x}(\cdot)$ and $\varepsilon = 0$ according to (2), which gives

$$\hat{J}_0 \leq J_0(\tilde{u}(\cdot)) \leq \liminf_{k\to\infty} J_{\varepsilon_k}(\tilde{u}_k(\cdot)) = \liminf_{k\to\infty} \hat{J}_{\varepsilon_k} , \qquad (3)$$

when the function $g(\cdot)$ is continuous.

Actually, the above question is related to the closeness of the set of trajectories in the space $C(R^n)$.We discuss this question in the next paragraph.

The chain of inequalities (3) gives us a lower bound for the optimal value.If we prove that

$$\limsup_{\varepsilon\to 0} \hat{J}_\varepsilon \leq \hat{J}_0 , \qquad (4)$$

the problem of performance well-posedness will be solved.Moreover, if we assume in addition that (1) has a solution $(\hat{u}_k(\cdot),\hat{x}_k(\cdot))$ for $\varepsilon = \varepsilon_k$,then the sequence $\hat{x}_k(\cdot)$ contains a subsequence,which converges uniformly to an optimal trajectory for the limit problem.

In order to prove (4) it is sufficient to find a sequence of controls $u_k(\cdot)$, feasible for the perturbed problem with ε_k , such that

$$\lim_{k\to\infty} J_{\varepsilon_k}(u_k(\cdot)) = \hat{J}_0 .$$

Then (4) follows from

$$J_{\varepsilon_k}(\tilde{u}_k(\cdot)) \leq J_{\varepsilon_k}(u_k(\cdot)) + \varepsilon_k .$$

In our case the situation is very simple.Let $u_k(\cdot)$ be a minimizing sequence for the basic problem, that is

$$J_0(u_k(\cdot)) \leq \hat{J}_0 + \varepsilon_k .$$

Assuming that the trajectory of the system is continuously depending on ε uniformly in u we have

$$\lim_{k\to\infty} (J_{\varepsilon_k}(u_k(\cdot)) - J_0(u_k(\cdot))) = 0 ,$$

which completes the proof of (4).

If, instead of (1) the performance index is in integral form

$$J(u(\cdot)) = \int_0^1 f(x(t),u(t),t)dt$$

and the system is linear with respect to the state,then our scheme can be essentially simplified.In this case it is convinient to choose a L^2 - weakly convergent subsequence of controls $\tilde{u}_k(\cdot)$.The boundedness of V can be replaced by a grouth condition for the function $f(\cdot)$, e.g.

$$f(x,u,t) \geqslant a|u|^2 \quad , a > 0 ,$$

for all $x \in R^n$,$u \in V$ and $t \in [0,1]$.It remains to assume (or prove in the concrete case) that the functional $J(u(\cdot))$ is L^2- weakly lower semicontinuous.Clearly, the admissible set of controls should be weakly closed (e.g. V is closed and convex).The upper estimate (4) will require some continuity conditions for the function $f(\cdot)$.This modification of the scheme is developed in Section 3.4 for a singularly perturbed problem.

The problem of well-posedness complcates considerably in presence of state constraints.For example, let us introduce an additional constraint

$$x(1) \in G ,$$

where G is a set in R^n.The lower bound (3) does not provide difficulties.Since $\tilde{x}_k(\cdot)$ is uniformly convergent to $\tilde{x}(\cdot)$,it is sufficient to take G closed.If G is perturbed, that is $G = G_\varepsilon$, the multifunction $G_{(\cdot)}$ should be upper semicontinuous.The upper bound (4),however,needs some more restrictive conditions.

Example 2.1.

$$- x(1) \longrightarrow \inf$$

$$\dot{x} = u - \varepsilon \quad , x(0) = 0,$$

$$-1 \leqslant u(t) \leqslant 1 ,$$

$$x(1) \in \{0,1\}.$$

The state constraining set is not convex, it consists of two points.For $\varepsilon = 0$ the optimal control $\hat{u}_0(t) = 1$ gives $\hat{J}_0 = -1$.If $\varepsilon > 0$ there is no feasible controls driving the state to 1, thus $\hat{J}_\varepsilon = 0$.

If the state constraining set is perturbed,e.g.

$$G_\varepsilon = \{x \in R^1 , |x-1| \leq \varepsilon \text{ or } |x| \leq \varepsilon\} ,$$

the optimal control $\hat{u}_\varepsilon(t)$ provides $\hat{J}_\varepsilon = -1 + \varepsilon$ and there is performance convergence.Notice that in this case $G_\varepsilon \supset G_o$ and G_ε tends to G_o in a rate proportional to the variation of the state due to change of ε in the differential equation.But if we take

$$G_\varepsilon = \{ x \in R^1 , |x-1| \leq \varepsilon^2 \text{ or } |x| \leq \varepsilon^2 \} ,$$

the problem will be not well-posed.

The above observation suggests that the state constraining set should vary but not too fast compared with the system dynamics.Such a requirement is discussed in the next paragraph.

One can avoid this assumption when the system is linear, the set G is convex and int $G \neq \emptyset$.The respective technique is presented in Section 3.4 for a singularly perturbed problem.

Observe that the presented scheme does not use any optimality conditions for the problems considered.In Section 2.3 we develop a different technique using the Lagrange duality theory for convex problems.

2.2.2. Relaxation and well-posedness

As before, let ε be a scalar perturbation parameter, $\varepsilon \in [0, \varepsilon_o]$, where ε_o will be chosen to be sufficiently small.For fixed ε consider the following optimal control problem, denoted as (P_ε):

$$J_\varepsilon(u(\cdot)) = g(x(1)) \longrightarrow \inf$$

subject to the constraints

$$\dot{x} = \varphi_\varepsilon(x,u,t) , \; x(0) = x^o , \tag{5}$$

$$x(t) \in G_\varepsilon(t) \subset R^n \quad \text{for all } t \in [0,1] , \tag{6}$$

$$u(t) \in V(t) \subset R^m \quad \text{for a.e. } t \in [0,1] , \tag{7}$$

$$x(\cdot) \in A(R^n) , \; u(\cdot) \in L^2(R^m) .$$

Denote by (P_0) the problem corresponding to $\varepsilon = 0$. We assume the following:

A1(i). The set $V(t)$ is compact for every $t \in [0,1]$ and the multifunction $V(\cdot): t \to V(t)$ is measurable. The set $G_\varepsilon(t)$ is closed for every $t \in [0,1]$ and $\varepsilon \in [0, \varepsilon_0]$ and upper semicontinuous with respect to ε for every $t \in [0,1]$.

(ii). There exists a compact set $X \subset R^n$ such that for every feasible control $u(\cdot)$ and for every $\varepsilon \in [0, \varepsilon_0]$ there exists unique solution $x_\varepsilon(\cdot)$ of the system (5) such that $x_\varepsilon(t) \in X$ for all $t \in [0,1]$.

(iii). The function $g(\cdot)$ is continuous. The function $\varphi_\varepsilon(\cdot)$ is a Caratheodory function with respect to $(x,u) \in X \times V(t)$ and $t \in [0,1]$ for every $\varepsilon \in [0, \varepsilon_0]$; moreover, for almost every $t \in [0,1]$ $\varphi_\varepsilon(x,u,t)$ is continuous with respect to x and ε on $X \times \{0\}$ uniformly in $u \in V(t)$.

(iv). Let $\hat{J}_0 = \inf J_0(u(\cdot)) < +\infty$ and $u_k(\cdot)$ be a minimizing sequence, that is

$$J_0(u_k(\cdot)) \leqslant \hat{J}_0 + \delta_k \text{ , } k = 1,2,\ldots \text{ ,}$$

where $\delta_k \to 0$ as $k \to \infty$. Let the trajectory $x_\varepsilon^k(\cdot)$ correspond to ε and $u_k(\cdot)$ and $x^k(\cdot)$ correspond to $\varepsilon = 0$ and $u_k(\cdot)$ according to (5). We suppose that for every $k = 1,2,\ldots$ there exists a function $\rho_k(\cdot)$, $\rho_k(\varepsilon) \to 0$ as $\varepsilon \to 0$, such that

$$\|x_\varepsilon^k - x^k\|_C \leqslant \rho_k(\varepsilon)$$

and

$$\{x \in R^n \text{ , } |x - x^k(t)| \leqslant \rho_k(\varepsilon)\} \subset G_\varepsilon(t)$$

for $k = 1,2,\ldots$, $t \in [0,1]$ and $\varepsilon \in [0, \varepsilon_0]$.

Along with the problem (P_0) we consider the following <u>relaxed</u> problem in the form of Warga [75]:

$$J^R(u(\cdot)) = g(x(1)) \to \inf$$

subject to

$$\dot{x} \in \operatorname{co} \varphi_0(x, V(t), t) \text{ , } x(0) = x^0, \tag{8}$$

$$x(t) \in G_0(t) \text{ , } t \in [0,1] \text{ ,}$$

where

$$\varphi_0(x,V(t),t) = \left\{\varphi_0(x,u,t) \ , \ u \in V(t)\right\} .$$

We say that the problem (P_0) is <u>regular in the sense of relaxation</u> if its optimal value is the same as that of the relaxed problem,i.e.

$$\hat{J}_0 = \hat{J}^R .$$

Observe that on our assumptions the relaxed problem has a solution and

$$\hat{J}^R \leqslant \hat{J}_0 .$$

The concept of relaxation has played a significant role in the development of the recent existence theory of optimal control, see Gamkrelidze [25], Eckeland and Temam [22], Warga [76], and the survey by Mordukhovič [55].It appears that the regularity in the sense of relaxation is a fairly common property of optimal control problems. For example, our problem (P_0) will be regular in the sense of relaxation when the system is linear with respect to the state,if the state is one-dimensional or the state constraints are vacuous.A number of sufficient conditions for regularity can be found in Warga [76], Joffe [41], Eckeland and Temam [22].

In the following result we develop the idea from Mordukhovič [56], where the concept of relaxation was applied to obtain well-posedness of a finite-difference approximation to nonconvex constrained optimal control problem.

<u>Theorem 2.1</u>. Suppose that the problem (P_0) is regular in the sense of relaxation.Then

$$\lim_{\varepsilon \to 0} \hat{J}_\varepsilon = \hat{J}_0 .$$

Proof. Applying the scheme from the previous paragraph we show that

$$\hat{J}^R \leqslant \liminf_{\varepsilon \to 0} \hat{J}_\varepsilon \leqslant \limsup_{\varepsilon \to 0} \hat{J}_\varepsilon \leqslant \hat{J}_0 \quad . \tag{9}$$

Then the regularity implies well-posedness.

Assume that there exists a sequence $\varepsilon_k \to 0$ such that

$$J^R > \lim_{k \to \infty} J_{\varepsilon_k} . \qquad (10)$$

Let $\tilde{u}_k(\cdot)$ be a sequence of suboptimal controls such that

$$J_{\varepsilon_k}(\tilde{u}_k(\cdot)) \leq \hat{J}_{\varepsilon_k} + \varepsilon_k , \; k = 1,2,\ldots ,$$

and $\tilde{x}_k(\cdot)$ be the corresponding sequence of the derivatives of the state.Clearly, the sequence $\tilde{x}_k(\cdot)$ has a L^2-weak limit point $p(\cdot)$, moreover, $\tilde{x}_k(\cdot)$ can be thoughtas uniformly convergent to

$$\tilde{x}(t) = x^0 + \int_0^t p(s)ds .$$

Let $[A]_\delta$ denote the closure of the δ-neighbourhood of the set A, that is

$$[A]_\delta = cl\{x \in R^n, \inf_{z \in A}|z - x| < \delta\}.$$

Then,choosing an arbitrary $\delta > 0$, from A1(iii) we conclude that for almost every $t \in [0,1]$

$$\dot{\tilde{x}}_k(t) = \varphi_{\varepsilon_k}(\tilde{x}_k(t), \tilde{u}_k(t), t) \in [\varphi_0(\tilde{x}(t), V(t), t)]_\delta$$

for large k.Applying Mazur's theorem we select a convex combination of $\dot{\tilde{x}}_k(\cdot)$ which converges strongly in $L^2(R^n)$ to $p(\cdot)$.Hence,it can be identified with a sequence which is pointwise convergent to $p(\cdot)$.In effect we get

$$p(t) \in \bigcap_{\delta > 0} co\,[\varphi_0(\tilde{x}(t), V(t), t)]_\delta .$$

In Mordukhovič [56] it is proved that

$$\bigcap_{\delta > 0} co\,[\varphi_0(x, V(t), t)]_\delta = co\; \varphi_0(x, V(t), t) .$$

Thus, $\tilde{x}(\cdot)$ is a trajectory of the relaxed system (8).Moreover, from the upper semicontinuity of $G_{(\cdot)}(t)$ we get that

$$\tilde{x}(t) \in G_0(t) \quad \text{for all } t \in [0,1] .$$

Hence

$$\lim_{k\to\infty} J_{\varepsilon_k}(\tilde{u}_k(\cdot)) = g(\tilde{x}(1)) \geqslant \hat{J}^R ,$$

which contradicts (10).

Now we prove that

$$\lim_{\varepsilon\to 0} \hat{J}_\varepsilon \leqslant \hat{J}_o . \tag{11}$$

On the contrary, there exists a sequence $\varepsilon_l \to 0$ as $l \to \infty$, such that

$$\lim_{l\to\infty} J_{\varepsilon_l} > \hat{J}_o .$$

Let us apply the minimizing sequence $u_k(\cdot)$ from A1(iv) to the perturbed system (5) with $\varepsilon = \varepsilon_l$. If we denote by $x^k_{\varepsilon_l}(\cdot)$ the corresponding solution, then

$$\| x^k_{\varepsilon_l} - x^k \|_C \leqslant \rho_k(\varepsilon_l) \quad , k=1,2,\ldots, \; l=1,2,\ldots,$$

where $x^k(\cdot)$ solves (5) for $\varepsilon = 0$ and $u_k(\cdot)$, and $\lim_{l\to\infty} \rho_k(\varepsilon_l) = 0$ for every $k = 1,2,\ldots$ Hence

$$x^k_{\varepsilon_l}(t) \in G_{\varepsilon_l}(t) \text{ for all } t\in[0,1] ,$$

i.e. $x^k_{\varepsilon_l}(\cdot)$ is an admissible trajectory for the perturbed problem.

This yields

$$g(x^k_{\varepsilon_l}(1)) \geqslant \hat{J}_{\varepsilon_l} .$$

Furthermore

$$\lim_{l\to\infty} g(x^k_{\varepsilon_l}(1)) \leqslant \hat{J}_o + \delta_k \quad , k = 1,2,\ldots$$

and for large k we come to a contradiction. This proves (11), Q.E.D.

<u>Remark 2.1</u>. The condition A1(iv) holds for <u>any</u> function $\rho_k(\cdot)$, determined by the system dynamics if $x_k(t) \in \operatorname{int} G_o(t)$ for all $t\in[0,1]$ and for some $\delta > 0$ the set

$$G_\varepsilon(t) \cap \{x \in R^n,\ |x - x_k(t)| < \delta\}$$

is continuous with respect to ε at $\varepsilon = 0$ for every $t \in [0,1]$.

Clearly, if φ is Lipschitz continuous with respect to x and ε with Lipschitz constants L_x and L_ε uniformly in u and t, one can take

$$\rho(\varepsilon) = L_\varepsilon \exp(L_x)\varepsilon \ .$$

Remark 2.2. We show that, on our assumptions, the regularity in the sense of relaxation is also a necessary condition for performance well-posedness. For simplicity, let $\rho(\varepsilon)$ in A1(iv) be $\rho(\varepsilon) = L\varepsilon$ and let $G_\varepsilon(t)$ contain a $(L+1)\varepsilon$-neighbourhood of $G_0(t)$ for every $t \in [0,1]$. We prove that

$$\lim_{\varepsilon \to 0} \hat{J}_\varepsilon = \hat{J}^R \ .$$

Then, if there is performance convergence, the problem (P_0) will be regular in the sense of relaxation.

Taking into account the first part of the proof of Theorem 2.1 it is sufficient to show that

$$\limsup_{\varepsilon \to 0} \hat{J}_\varepsilon \leqslant \hat{J}^R \ .$$

Assume the opposite. Then there exists a sequence $\varepsilon_k \to 0$ as $k \to \infty$ such that

$$\lim_{k \to \infty} \hat{J}_{\varepsilon_k} > \hat{J}^R \ . \tag{12}$$

Let $\hat{x}(\cdot)$ be an optimal trajectory of the relaxed problem. From Warga [75], there exists a sequence of trajectories $x^l(\cdot)$ of the system (5) with $\varepsilon = 0$, corresponding to controls $u^l(\cdot)$, $u^l(t) \in V(t)$ for a.e. $t \in [0,1]$, such that

$$\|x^l - \hat{x}\|_C = \delta_l \to 0 \text{ as } l \to \infty \ .$$

Let us apply $u^l(\cdot)$ to the perturbed system (5) with ε_k. The resulting trajectory fulfilles

$$\|x^1_{\varepsilon_k} - \hat{x}\|_C \leq \delta_1 + L\varepsilon_k \quad . \tag{13}$$

For every k we choose δ_{1_k} so small that $\delta_{1_k} < \varepsilon_k$.Then

$$x^{1_k}_{\varepsilon_k}(t) \in G_{\varepsilon_k}(t) \text{ for all } t \in [0,1]$$

and

$$g(x^{1_k}_{\varepsilon_k}(1)) \geq \hat{J}_{\varepsilon_k} \quad ,$$

which,combined with (13), contradicts (12), Q.E.D.

In this chapter we shall not discuss the question of well-posedness in the sense of solution convergence.Some results concerning the convergence of the optimal controls for singularly perturbed problems are given in Chapter 3.The next section presents results in a different direction: we give estimations of the changes of the optimal control due to regular deviation of the data.

2.3.Estimates of the optimal solution for convex constrained problems

In this section we shall consider convex control problems with state and control constraints.We study first the following problem

$$I_\varepsilon(x(\cdot),u(\cdot)) = \int_0^1 f_\varepsilon(x(t),u(t),t)dt \longrightarrow \inf \qquad (CP_\varepsilon)$$

$$\dot{x} = A_\varepsilon(t)x + B_\varepsilon(t)u \quad , \; x(0) = x^0 \quad , \; t \in [0,1] \quad , \tag{14}$$

$$x(\cdot) \in X_\varepsilon \subset H^1(R^n) \quad , \; u(\cdot) \in U_\varepsilon \subset L^2(R^m) \quad , \tag{15}$$

where ε ,as before, represents the perturbation, $\varepsilon \in [0,\varepsilon_0]$ .The basic problem (CP_0) corresponds to $\varepsilon = 0$.

Remark 2.3. In all further proofs and statements we designate by c a generic constant, which does not depend on the perturbation ε and on the time t but may change in different relations.

2.3.1.Set constraints

We assume the following:

A2(i). The elements of the matrices $A_\varepsilon(t)$ and $B_\varepsilon(t)$ are bounded on $[0,1]\times[0,\varepsilon_0]$,square integrable with respect to t on $[0,1]$ and Lipschitz continuous with respect to ε at $\varepsilon = 0$ with Lipschitz constants in $L^2(R^1)$.

(ii). The function $f_\varepsilon(x,u,t)$ and the derivatives

$$\frac{\partial}{\partial(x,u)} f_\varepsilon(x,u,t) \ , \quad \frac{\partial^2}{\partial(x,u)^2} f_\varepsilon(x,u,t)$$

are continuous on $R^{n+m}\times[0,1]\times[0,\varepsilon_0]$.There exists $\alpha > 0$ such that for all $y = (x,u) \in R^{n+m}$, $z \in R^{n+m}$ and $t \in [0,1]$

$$z^T \frac{\partial^2}{\partial y^2} f_0(y,t)z \geqslant \alpha|u|^2 \quad .$$

(iii). The sets X_ε and U_ε are closed and convex for every $\varepsilon \in [0,\varepsilon_0]$. There exists a continuous control $\bar{u}_\varepsilon(\cdot) \in U_\varepsilon$ for every $\varepsilon \in [0,\varepsilon_0]$ such that $\limsup_{\varepsilon \to 0} \|\bar{u}_\varepsilon\|_C < +\infty$ and the corresponding trajectory $\bar{x}_\varepsilon(\cdot) \in \operatorname{int} X_\varepsilon$ for $\varepsilon \in [0,\varepsilon_0]$.

The condition A2(ii) means that the functional $I_\varepsilon(x(\cdot),u(\cdot))$ is convex with respect to $(x(\cdot),u(\cdot))$ and strongly convex with respect to $u(\cdot)$.The condition A2(iii) is related to the known Slater's condition.

By a standard argument,see Joffe and Tikhomirov [42] ,p.368 , for $\varepsilon = 0$ and for ε sufficiently small there exist unique solutions $(\hat{x}_0(\cdot),\hat{u}_0(\cdot))$ and $(\hat{x}_\varepsilon(\cdot),\hat{u}_\varepsilon(\cdot))$ of the problems (CP_0) and (CP_ε) respectively.Along with A2(i)-(iii) we suppose that

A2(iv). The function

$$\frac{\partial}{\partial(x,u)} f_\varepsilon(\hat{x}_0(t),\hat{u}_0(t),t)$$

is Lipschitz continuous with respect to ε at $\varepsilon = 0$ with a Lipschitz constant in $L^2(R^1)$.

Using Hahn-Banach theorem as in Hager and Mitter [39] we conclude that there exists a function $p_\varepsilon(\cdot) \in L^2(R^n)$ such that for every $x(\cdot)$ in X_ε and $u(\cdot)$ in U_ε

$$\hat{I}_\varepsilon = I_\varepsilon(\hat{x}_\varepsilon(\cdot),\hat{u}_\varepsilon(\cdot)) \leqslant I_\varepsilon(x(\cdot),u(\cdot)) + \langle p_\varepsilon, \dot{x} - A_\varepsilon x - B_\varepsilon u\rangle.$$

The following assumption is related to the optimality conditions for convex problems:

A3. There exists a function $p_0(\cdot) \in L^2(R^n)$ such that for every $x(\cdot) \in X_0$ and $u(\cdot) \in U_0$

$$\langle \frac{\partial}{\partial x} f_0(\hat{x}_0,\hat{u}_0) - A_0^T p_0, x - \hat{x}_0\rangle + \langle p_0, \dot{x} - \dot{\hat{x}}_0\rangle \geqslant 0 \quad ,$$

$$\langle \frac{\partial}{\partial u} f_0(\hat{x}_0,\hat{u}_0) - B_0^T p_0, u - \hat{u}_0\rangle \geqslant 0 \ .$$

The function $p_0(\cdot)$ corresponds to the adjoint variable in the Maximum principle for the problem (CP_0). Sufficient conditions for existence of $p_0(\cdot)$ can be found in Girsanov [31], Hager and Mitter [39] and Rockafellar [65].

In the sequel we assume that ε is chosen to be sufficiently small.

Lemma 2.1.

$$\|\hat{x}_\varepsilon - \hat{x}_0\|_C + \|\hat{x}_\varepsilon - \hat{x}_0\|_H \leqslant c(\|\hat{u}_\varepsilon - \hat{u}_0\| + \varepsilon) \ . \qquad (16)$$

This inequality follows directly from the linearity of the system, A2(i) and the Gronwall lemma.

Lemma 2.2.

$$\limsup_{\varepsilon \to 0} (\|\hat{u}_\varepsilon\| + \hat{I}_\varepsilon + \|p_\varepsilon\|) < +\infty \ .$$

Proof. The strong convexity assumption A2(ii) implies

$$I_\varepsilon(\bar{x}_\varepsilon(\cdot),\bar{u}_\varepsilon(\cdot)) - \hat{I}_\varepsilon \geqslant \alpha\|\bar{u}_\varepsilon - \hat{u}_\varepsilon\|^2 . \qquad (17)$$

Furthermore

$$I_\varepsilon(\bar{x}_\varepsilon(\cdot),\bar{u}_\varepsilon(\cdot)) - \hat{I}_\varepsilon \leqslant \langle \frac{\partial}{\partial x} f_\varepsilon(\bar{x}_\varepsilon,\bar{u}_\varepsilon), \bar{x}_\varepsilon - \hat{x}_\varepsilon\rangle + \langle \frac{\partial}{\partial u} f_\varepsilon(\bar{x}_\varepsilon,\bar{u}_\varepsilon), \bar{u}_\varepsilon - \hat{u}_\varepsilon\rangle \ . \qquad (18)$$

Subtracting (17) from (18) and using (16) we obtain that $\|\hat{u}_\varepsilon\|$ and $\hat{I}_\varepsilon$ are bounded when $\varepsilon \to 0$.

Assume that $\lim_{\varepsilon\to 0}\|p_\varepsilon\| = +\infty$ .Let $\Delta x_\varepsilon(\cdot)$ be the solution of the equation

$$\Delta\dot{x} = A_\varepsilon(t)\Delta x - p_\varepsilon(t)/\|p_\varepsilon\| , \qquad \Delta x(0) = 0.$$

Then for sufficiently small but positive δ the trajectory $\tilde{x}_\varepsilon(\cdot) = \bar{x}_\varepsilon(\cdot) + \delta\Delta x_\varepsilon(\cdot) \in X_\varepsilon$.Since $\limsup_{\varepsilon\to 0}\|\bar{x}_\varepsilon\|_C < +\infty$ we get

$$\limsup_{\varepsilon\to 0}|I_\varepsilon(\tilde{x}_\varepsilon(\cdot),\bar{u}_\varepsilon(\cdot))| < +\infty .$$

Hence

$$\hat{I}_\varepsilon - I_\varepsilon(\tilde{x}_\varepsilon(\cdot),\bar{u}_\varepsilon(\cdot)) \leqslant \langle p_\varepsilon,\dot{\tilde{x}}_\varepsilon - A_\varepsilon\tilde{x}_\varepsilon - B_\varepsilon\bar{u}_\varepsilon\rangle = -\delta\|p_\varepsilon\| < 0 .$$

We come to a contradiction.Hence $\|p_\varepsilon\|$ is bounded,Q.E.D.

Let $(x_0^\varepsilon(\cdot),u_0^\varepsilon(\cdot))$ be the projection of $(\hat{x}_0(\cdot),\hat{u}_0(\cdot))$ at $X_\varepsilon\times U_\varepsilon$ and $(x_\varepsilon(\cdot),u_\varepsilon(\cdot))$ be the projection of $(\hat{x}_\varepsilon(\cdot),\hat{u}_\varepsilon(\cdot))$ at $X_0\times U_0$.Clearly, $\|x_0^\varepsilon\|_H$ and $\|u_0^\varepsilon\|$ are bounded when $\varepsilon\to 0$.Define the number

$$d(\varepsilon) = \|x_0^\varepsilon - \hat{x}_0\|_H + \|x_\varepsilon - \hat{x}_\varepsilon\|_H + \|u_0^\varepsilon - \hat{u}_0\| + \|u_\varepsilon - \hat{u}_\varepsilon\| .$$

<u>Theorem 2.2</u>. Suppose that

A4. $$\limsup_{\varepsilon\to 0}\left(\left\|\frac{\partial}{\partial x}f_\varepsilon(x_0^\varepsilon,u_0^\varepsilon)\right\|_{L^1} + \left\|\frac{\partial}{\partial u}f_\varepsilon(x_0^\varepsilon,u_0^\varepsilon)\right\|\right) < +\infty .$$

Then

$$\|\hat{u}_\varepsilon - \hat{u}_0\| \leqslant c(\varepsilon + d(\varepsilon))^{0.5} .$$

Proof. We have

$$\hat{I}_\varepsilon \leqslant I_\varepsilon(x_0^\varepsilon(\cdot),u_0^\varepsilon(\cdot)) + \langle p_\varepsilon,\dot{x}_0^\varepsilon - A_\varepsilon x_0^\varepsilon - B_\varepsilon u_0^\varepsilon\rangle$$

$$\leqslant I_\varepsilon(\hat{x}_0(\cdot),\hat{u}_0(\cdot)) + \langle\frac{\partial}{\partial x}f_\varepsilon(x_0^\varepsilon,u_0^\varepsilon),x_0^\varepsilon - \hat{x}_0\rangle$$

$$+ \langle\frac{\partial}{\partial u}f_\varepsilon(x_0^\varepsilon,u_0^\varepsilon),u_0^\varepsilon - \hat{u}_0\rangle$$

$$+ \langle p_\varepsilon,\dot{x}_0^\varepsilon - \dot{\hat{x}}_0 - A_0(x_0^\varepsilon - \hat{x}_0) - B_0(u_0^\varepsilon - \hat{u}_0)$$

$$- (A_\varepsilon - A_o)x_o^\varepsilon - (B_\varepsilon - B_o)u_o^\varepsilon >$$

$$\leqslant I_\varepsilon(\hat{x}_o(\cdot),\hat{u}_o(\cdot)) + c(\varepsilon + d(\varepsilon)) . \qquad (19)$$

Furthermore, by A2(ii) and A3

$$\hat{I}_\varepsilon \geqslant I_\varepsilon(\hat{x}_o(\cdot),\hat{u}_o(\cdot)) + \langle p_\varepsilon, \dot{\hat{x}}_o - A_\varepsilon\hat{x}_o - B_\varepsilon\hat{u}_o\rangle$$
$$+ \langle \frac{\partial}{\partial x}(f_\varepsilon(\hat{x}_o,\hat{u}_o) - f_o(\hat{x}_o,\hat{u}_o)), \hat{x}_\varepsilon - \hat{x}_o\rangle$$
$$+ \langle \frac{\partial}{\partial u}(f_\varepsilon(\hat{x}_o,\hat{u}_o) - f_o(\hat{x}_o,\hat{u}_o)), \hat{u}_\varepsilon - \hat{u}_o\rangle$$
$$+ \langle (A_\varepsilon - A_o)^T p_o, \hat{x}_\varepsilon - \hat{x}_o \rangle + \langle (B_\varepsilon - B_o)^T p_o, \hat{u}_\varepsilon - \hat{u}_o\rangle$$
$$+ \langle \frac{\partial}{\partial x} f_o(\hat{x}_o,\hat{u}_o) - A_o{}^T p_o, \hat{x}_\varepsilon - x_\varepsilon\rangle + \langle p_o, \dot{\hat{x}}_\varepsilon - \dot{x}_\varepsilon \rangle$$
$$+ \langle \frac{\partial}{\partial u} f_o(\hat{x}_o,\hat{u}_o) - B_o{}^T p_o, \hat{u}_\varepsilon - u_\varepsilon\rangle + \alpha\|\hat{u}_\varepsilon - \hat{u}_o\|^2 .$$

The last inequality yields

$$\hat{I}_\varepsilon \geqslant I_\varepsilon(\hat{x}_o(\cdot),\hat{u}_o(\cdot)) + \alpha\|\hat{u}_\varepsilon - \hat{u}_o\|^2 - c(d(\varepsilon) + \varepsilon\|\hat{u}_\varepsilon - \hat{u}_o\|). \quad (20)$$

Subtracting (19) from (20) we get

$$\|\hat{u}_\varepsilon - \hat{u}_o\|^2 \leqslant c(\varepsilon\|\hat{u}_\varepsilon - \hat{u}_o\| + d(\varepsilon) + \varepsilon \quad) ,$$

which gives us the desired estimate.

Notice that $d(\varepsilon)$ can be estimated by the sum of the Hausdorff distances between the sets X_ε, X_o and U_ε, U_o.

Remark 2.4. From (19) and (20) we immediately obtain performance well-posedness; moreover, for that purpose we need some convexity and continuity conditions weaker than A2. Furthermore, an uniform convexity assumption will imply strong convergence of the optimal controls.

2.3.2. Inequality constraints

We continue the analysis of the problem (CP_ε) assuming that the sets X_ε and U_ε are defined by local inequalities, that is

$$X_\varepsilon = \{ x(\cdot) \in A(R^n),\ \Theta_\varepsilon(x(t)t) \leq 0 \text{ for all } t \in [0,1] \},$$

$$U_\varepsilon = \{ u(\cdot) \in L^2(R^m),\ \Psi_\varepsilon(u(t),t) \leq 0 \text{ for a.e. } t \in [0,1] \},$$

where $\Theta_\varepsilon: R^n \times [0,1] \to R^p$, $\Psi_\varepsilon : R^m \times [0,1] \to R^q$. Let A2 and the following conditions hold:

A5. The functions $\Theta_\varepsilon(x,t), \Psi_\varepsilon(u,t)$ are continuous with respect to all the arguments; $\Theta_\varepsilon(\cdot,t), \Psi_\varepsilon(\cdot,t)$ are convex for all $\varepsilon \in [0,\varepsilon_0]$ and $t \in [0,1]$; $\Theta_\varepsilon(\cdot,\cdot)$ is C^2 and all the derivatives are continuous; $\Psi_\varepsilon(\cdot,\cdot)$ is C^1 and its derivatives are continuous; $\Theta_\varepsilon(\hat{x}_0(t),t), \Psi_\varepsilon(\hat{u}_0(t),t)$ and $\frac{\partial}{\partial x}\Theta_\varepsilon(\hat{x}_0(t),t)$ are Lipschitz continuous with respect to ε at $\varepsilon = 0$ uniformly in t; $\frac{\partial}{\partial u}\Psi_\varepsilon(\hat{u}_0(t),t)$ is Lipschitz continuous with respect to ε at $\varepsilon = 0$ with a Lipschitz constant in $L^2(R^1)$; there exists a constant $\beta < 0$ and a continuous control $\bar{u}(\cdot)$ such that

$$\Theta_0(\bar{x}(t),t)_i \leq \beta, \quad \Psi_0(\bar{u}(t),t)_j \leq \beta$$

for all $i=1,\ldots,p$, $j=1,\ldots,q$, $t \in [0,1]$, where $\bar{x}(\cdot)$ corresponds to $\bar{u}(\cdot)$ and $\varepsilon = 0$ according to (14).

Following Hager and Mitter [39] we introduce Lagrange functional in the form

$$L_\varepsilon(x(\cdot),u(\cdot);p(\cdot),\nu(\cdot),\lambda(\cdot)) = I_\varepsilon(x(\cdot),u(\cdot))$$
$$+ \langle p,\dot{x} - A_\varepsilon x - B_\varepsilon u\rangle + [\nu,\Theta_\varepsilon(x)] + \langle \lambda,\Psi_\varepsilon(u)\rangle ,$$

where $p(\cdot) \in BV(R^n)$, $\nu(\cdot) \in BV(R^p)$, $\nu(\cdot)$ is nondecreasing and $\nu(1) = 0$, $\lambda(\cdot) \in L^1(R^q)$, $\lambda(\cdot) \geq 0$, and

$$[f,g] \quad \text{denotes} \quad \int_0^1 g^T(t)df(t) .$$

The next theorem is proved in Hager and Mitter [39] (see also Hager [38]).

<u>Theorem 2.3</u>. There exist Lagrange multipliers $p_\varepsilon(\cdot)$, $\nu_\varepsilon(\cdot)$, $\lambda_\varepsilon(\cdot)$ such that

$$\hat{I}_\varepsilon = \min L_\varepsilon(x(\cdot),u(\cdot);p_\varepsilon(\cdot),\nu_\varepsilon(\cdot),\lambda_\varepsilon(\cdot)) , \quad x(\cdot) \in A(R^n),$$
$$x(0) = x^0 , \quad u(\cdot) \in L^2(R^m) .$$

If $q_\varepsilon(t) = G_\varepsilon^T(t)\nu_\varepsilon(t) - p_\varepsilon(t)$, where $G_\varepsilon(t) = \frac{\partial}{\partial x}\theta_\varepsilon(\hat{x}_\varepsilon(t),t)$, then $q_\varepsilon(\cdot) \in A(R^n)$, $q_\varepsilon(1) = 0$, and the following relations hold:

$$\frac{\partial}{\partial u}f_\varepsilon(\hat{x}_\varepsilon(t),\hat{u}_\varepsilon(t),t) - B_\varepsilon^T(t)(G_\varepsilon^T(t)\nu_\varepsilon(t) - q_\varepsilon(t)) + \frac{\partial}{\partial u}\psi_\varepsilon^T(\hat{u}_\varepsilon(t),t)\lambda_\varepsilon(t) = 0 , \tag{21}$$

$$\dot{q}_\varepsilon(t) = - A_\varepsilon^T(t)(q_\varepsilon(t) - G_\varepsilon^T(t)\nu_\varepsilon(t)) - \frac{\partial}{\partial x}f_\varepsilon(\hat{x}_\varepsilon(t),\hat{u}_\varepsilon(t),t) + \frac{d}{dt}G_\varepsilon^T(t)\nu_\varepsilon(t) , \tag{22}$$

$$[\nu_\varepsilon,\Theta_\varepsilon(\hat{x}_\varepsilon)] = 0 \quad , \quad \langle\lambda_\varepsilon,\Psi_\varepsilon(\hat{u}_\varepsilon)\rangle = 0 . \tag{23}$$

Lemma 2.3.

$$\limsup_{\varepsilon\to 0} (\|\hat{u}_\varepsilon\|_{L^\infty} + \|p_\varepsilon\|_C + V_0^1(\nu_\varepsilon) + \|\lambda_\varepsilon\|_{L^1}) < +\infty .$$

Proof. From Lemma 2.2 we have

$$\limsup_{\varepsilon\to 0} (\|\hat{u}_\varepsilon\| + \|\hat{x}_\varepsilon\|_C + \hat{I}_\varepsilon) < +\infty .$$

Since

$$\hat{I}_\varepsilon \leqslant L_\varepsilon(\bar{x}(\cdot),\bar{u}(\cdot);p_\varepsilon(\cdot),\nu_\varepsilon(\cdot),\lambda_\varepsilon(\cdot))$$

for sufficiently small ε we get

$$\hat{I}_\varepsilon - I_\varepsilon(\bar{x}(\cdot),\bar{u}(\cdot)) \leqslant [\nu_\varepsilon,\Theta_\varepsilon(\bar{x})] + \langle\lambda_\varepsilon,\Psi_\varepsilon(\bar{u})\rangle$$

$$\leqslant \frac{\beta}{2}\Big(\int_0^1 \sum_{i=1}^{q} \lambda_\varepsilon(t)_i dt + \int_0^1 \sum_{i=1}^{p} d\nu_\varepsilon(t)_i \Big) \leqslant 0.$$

Hence

$$\limsup_{\varepsilon\to 0} (V_0^1(\nu_\varepsilon) + \|\lambda_\varepsilon\|_{L^1}) < +\infty .$$

Applying Gronwall lemma to (22) we deduce

$$\limsup_{\varepsilon\to 0} \|q_\varepsilon\|_C < +\infty .$$

Then $\limsup_{\varepsilon \to 0} \|p_\varepsilon\|_C < +\infty$ .Introduce the Hamiltonian

$$H_\varepsilon(u,t) = - f_\varepsilon(\hat{x}_\varepsilon(t),u,t) + p_\varepsilon^T(t)B_\varepsilon(t)u .$$

If one defines

$$W_\varepsilon(t) = \{ u \in R^m , \Psi_\varepsilon(u,t) \leqslant 0 \} ,$$

then for almost all $t \in [0,1]$

$$H_\varepsilon(\hat{u}_\varepsilon(t),t) = \max H_\varepsilon(u,t) \quad , \ u \in W_\varepsilon(t) .$$

The strong concavity of $H(\cdot,t)$ yields

$$\alpha|\hat{u}_\varepsilon(t) - \bar{u}(t)| \leqslant |\frac{\partial}{\partial u} H_\varepsilon(\bar{u}(t),t)|$$

for almost all $t \in [0,1]$ .Hence $\|\hat{u}_\varepsilon\|_{L^\infty}$ is bounded when $\varepsilon \to 0$,Q.E.D.

Theorem 2.4.

$$\|\hat{u}_\varepsilon - \hat{u}_0\| \leqslant c\sqrt{\varepsilon} . \tag{24}$$

Proof.We have

$$\begin{aligned}\hat{I}_\varepsilon &\leqslant L_\varepsilon(\hat{x}_0(\cdot),\hat{u}_0(\cdot);p_\varepsilon(\cdot),\nu_\varepsilon(\cdot),\lambda_\varepsilon(\cdot)) \\ &\leqslant I_\varepsilon(\hat{x}_0(\cdot),\hat{u}_0(\cdot)) + \langle p_\varepsilon,(A_0 - A_\varepsilon)\hat{x}_0 + (B_0 - B_\varepsilon)\hat{u}_0 \rangle \\ &\quad + [\nu_\varepsilon, \Theta_\varepsilon(\hat{x}_0) - \Theta_0(\hat{x}_0)] + \langle \lambda_\varepsilon, \Psi_\varepsilon(\hat{u}_0) - \Psi_0(\hat{u}_0)\rangle . \end{aligned} \tag{25}$$

On the other hand, by A2(ii) and (23) we get

$$\begin{aligned}\hat{I}_\varepsilon &\geqslant \hat{I}_\varepsilon + \langle p_0, \hat{\dot{x}}_\varepsilon - A_\varepsilon\hat{x}_\varepsilon - B_\varepsilon\hat{u}_\varepsilon\rangle + [\nu_0, \Theta_\varepsilon(\hat{x}_\varepsilon)] + \langle \lambda_0, \Psi_\varepsilon(\hat{u}_\varepsilon)\rangle \\ &\geqslant I_\varepsilon(\hat{x}_0(\cdot),\hat{u}_0(\cdot)) + \langle p_0,(A_0 - A_\varepsilon)\hat{x}_0 + (B_0 - B_\varepsilon)\hat{u}_0\rangle \\ &\quad + [\nu_0, \Theta_\varepsilon(\hat{x}_0) - \Theta_0(\hat{x}_0)] + \langle \lambda_0, \Psi_\varepsilon(\hat{u}_0) - \Psi_0(\hat{u}_0)\rangle \\ &\quad + \langle \frac{\partial}{\partial u} f_\varepsilon(\hat{x}_0,\hat{u}_0) - B_\varepsilon^T p_0 + \frac{\partial}{\partial u}\Psi_\varepsilon^T(\hat{u}_0)\lambda_0 , \hat{u}_\varepsilon - \hat{u}_0 \rangle \\ &\quad + \langle \frac{\partial}{\partial x} f_\varepsilon(\hat{x}_0,\hat{u}_0) - A_\varepsilon^T p_0, \hat{x}_\varepsilon - \hat{x}_0 \rangle + \langle p_0, \hat{\dot{x}}_\varepsilon - \hat{\dot{x}}_0 \rangle \end{aligned}$$

$$+ \quad [\nu_o, \frac{\partial}{\partial x}\Theta_\varepsilon(\hat{x}_o)(\hat{x}_\varepsilon - \hat{x}_o)] \qquad + \alpha\|\hat{u}_\varepsilon - \hat{u}_o\|^2 . \tag{26}$$

Since $p_o(\cdot)$ is continuous from the left,an integration by parts gives us

$$\langle p_o,\hat{x}_\varepsilon - \hat{x}_o\rangle = p(1^-)^T(\hat{x}_\varepsilon(1) - \hat{x}_o(1)) - \int_0^1(\hat{x}_\varepsilon(t) - \hat{x}_o(t))^T dp_o(t)$$

$$= \int_0^1(\hat{x}_\varepsilon(t) - \hat{x}_o(t))^T(\dot{q}_o(t) - \dot{G}_o^T(t)\nu_o(t))dt$$

$$+ \nu_o(1^-)^T G_o(1)(\hat{x}_\varepsilon(1) - \hat{x}_o(1))$$

$$- \int_0^1(\hat{x}_\varepsilon(t) - \hat{x}_o(t))^T G_o^T(t)d\nu_o(t) . \tag{27}$$

Furthermore

$$[\nu_o, \frac{\partial}{\partial x}\Theta_\varepsilon(\hat{x}_o)(\hat{x}_\varepsilon - \hat{x}_o)] = -\nu_o(1^-)^T \frac{\partial}{\partial x}\Theta_\varepsilon(\hat{x}_o(1),1)(\hat{x}_\varepsilon(1)$$

$$-\hat{x}_o(1)) - \int_0^1(\hat{x}_\varepsilon(t) - \hat{x}_o(t))^T \frac{\partial}{\partial x}\Theta_\varepsilon(\hat{x}_o(t),t)d\nu_o(t) . \tag{28}$$

Substituting (27) and (28) in (26), using the relations (21),(22) (23) and combining (25) and (26) we obtain finally

$$\|\hat{u}_\varepsilon - \hat{u}_o\|^2 \leqslant c(\varepsilon\|\hat{u}_\varepsilon - \hat{u}_o\| + \varepsilon\|\hat{x}_\varepsilon - \hat{x}_o\| + \|p_\varepsilon - p_o\|$$

$$+ [\nu_\varepsilon - \nu_o,\Theta_\varepsilon(\hat{x}_o) - \Theta_o(\hat{x}_o)]$$

$$+ \langle\lambda_\varepsilon - \lambda_o,\Psi_\varepsilon(\hat{u}_o) - \Psi_o(\hat{u}_o)\rangle) . \tag{29}$$

Thus, by lemmas 2.1 and 2.3 we conclude that (26) holds,Q.E.D.

2.3.3.Regular inequality constraints

Since the data of the problem considered are Lipschitz continuous with respect to ε , the following question arises: under what requirement the optimal control is also Lipschitz continuous at $\varepsilon = 0$?

Observe that, taking into account (29), for such a result we need an estimate of the changes of the dual variables with respect to the control variations.

We take up again the problem (CP_ε) under the conditions A2,A5 and

A6. The matrices $A_\varepsilon(t)$ and $B_\varepsilon(t)$ are continuous with respect to t and Lipschitz continuous with respect to ε at $\varepsilon = 0$ uniformly in $t \in [0,1]$. The function $f_\varepsilon(\cdot)$ is C^2; the derivative

$$\frac{\partial}{\partial(x,u)} f_\varepsilon(\hat{x}_0(t),\hat{u}_0(t),t)$$

is Lipschitz continuous with respect to ε at $\varepsilon = 0$ uniformly in $t \in [0,1]$. The function $\Theta_\varepsilon(\cdot)$ is C^3, $\Psi_\varepsilon(\cdot)$ is C^2; $\Theta_\varepsilon(\hat{x}_0(t),t)$, $\Psi_\varepsilon(\hat{u}_0(t),t)$ and their derivatives at $(\hat{x}_0(t),\hat{u}_0(t),t)$ are Lipschitz continuous with respect to ε at $\varepsilon = 0$ uniformly in $t \in [0,1]$. The function $\Theta_\varepsilon(x^0,0)$ is twice continuously differentiable with respect to ε at $\varepsilon = 0$ (extending the domain of $\Theta_\varepsilon(x^0,0)$ on a neighbourhood of $\varepsilon = 0$).

A7. There exists $\gamma > 0$ such that for all $t \in [0,1]$ and z

$$\left|[G_c^T(t),B_0^T(t)G_s^T(t)]\, z\right| \geqslant \gamma |z| ,$$

where $G_c(t)$ is the matrix which rows are the derivatives at $\hat{u}_0(t)$ of the components of $\Psi_0(u,t)$ with respect to u, corresponding to the binding constraints for $\hat{u}_0(t)$. The matrix $G_s(t)$ is defined similarly from $\Theta_0(x,t)$.

The condition A7 was introduced by Hager [38], who proved the following:

<u>Theorem 2.5</u>. There exists an optimal solution $(\hat{x}_0(\cdot),\hat{u}_0(\cdot))$ of the problem (CP_0) and dual multipliers $p_0(\cdot), q_0(\cdot)$, $\nu_0(\cdot)$, $\lambda_0(\cdot)$ such that $\hat{x}_0(\cdot),\hat{u}_0(\cdot),p_0(\cdot),\dot{q}_0(\cdot)$, $\nu_0(\cdot)$, $\lambda_0(\cdot)$ are continuous with respect to t on [0,1).

In this paragraph we prove that

<u>Theorem 2.6</u>. The following estimation holds

$$\|\hat{u}_\varepsilon - \hat{u}_0\| \leqslant c\varepsilon .$$

Moreover, as a corollary we obtain similar estimations for the state and for the dual variables.

The proof is presented as a series of lemmas.

Lemma 2.4.

$$[\nu_\varepsilon - \nu_0, \Theta_\varepsilon(\hat{x}_0) - \Theta_0(\hat{x}_0)] \leqslant c\varepsilon(\|\nu_\varepsilon - \nu_0\|_{L^1} + \varepsilon).$$

Proof. Since $\Theta_0(x^0,0) \leqslant \beta$ there exists $t'>0$ such that

$$\Theta_\varepsilon(\hat{x}_\varepsilon(t),t) \leqslant \beta/2$$

for all $t \in [0,t']$ and ε sufficiently small, see Theorem 2.4 and Lemma 2.1. Hence $\nu_\varepsilon(t)$ and $\nu_0(t)$ are constants on $[0,t']$. Choose a function $w(\cdot) \in C^1(R^p)$, $w(t) = 0$ for $t \geqslant t'$, such that

$$w(0) = -\frac{\partial}{\partial\varepsilon}\Theta_0(x^0,0) .$$

Then $[\nu_\varepsilon - \nu_0, w] = 0$. Denote $\Delta\nu = \nu_\varepsilon - \nu_0$, $\Delta\Theta(t) = \Theta_\varepsilon(\hat{x}_0(t),t) - \Theta_0(x_0(t),t)$. We have

$$\begin{aligned}[\Delta\nu,\Delta\Theta] &= [\Delta\nu, \Delta\Theta + \varepsilon w] = -\Delta\Theta^T(1)\Delta\nu(1^-) \\ &\quad + \int_0^1 (\Delta\Theta(t) - \varepsilon w(t))^T d\Delta\nu(t) \\ &= -(\Delta\Theta(0) - \varepsilon w(0))^T\Delta\nu(0) - \int_0^1 (\Delta\dot{\Theta}(t) - \varepsilon\dot{w}(t))^T\Delta\nu(t)dt .\end{aligned}$$

Since

$$\Delta\Theta(0) + \varepsilon w(0) = O(\varepsilon^2)$$

and $\Delta\nu(\cdot)$ is uniformly bounded (Lemma 2.3), the proof is complete.

The following lemma shows that Hager's condition A7 can be extended to the sets of "almost binding" constraints. This lemma was proved first in Dontchev [14]. W.W. Hager proposed a shorter proof, which we present here.

Lemma 2.5. There exists $\delta_0 > 0$ such that for every $\delta \in (0,\delta_0)$ there exists $\gamma(\delta) > 0$, $\gamma(\delta) \to 0$ when $\delta \to 0$, so that for every $t \in [0,1]$ and $z = (z^1,z^2)$

$$\Big| \sum_{j\in c_\delta(t)} \frac{\partial}{\partial u}\Psi_0(\hat{u}_0(t),t)_j z_j^1 + \sum_{j\in r_\delta(t)} (B_0^T(t)\frac{\partial}{\partial x}\Theta_0(\hat{x}_0(t),t))_j z_j^2 \Big|$$

$$\geq \gamma(\delta)(|z^1| + |z^2|) ,$$

where the sets of indices $c_\delta(t)$ and $r_\delta(t)$ are defined as follows

$$c_\delta(t) = \{ j \in \{1,\dots,q\},\ \Psi_o(\hat{u}_o(t),t)_j \geq -\delta \} ,$$

$$r_\delta(t) = \{ j \in \{1,\dots,p\},\ \Theta_o(\hat{x}_o(t),t)_j \geq -\delta \} .$$

Proof.For every $t \in [0,1]$ there exists an open interval Δ centered at t, such that the constraints nonbinding at t remain nonbinding over $\mathrm{cl}\,\Delta$.This follows from the continuity of $\hat{x}_o(t)$ and $\hat{u}_o(t)$. Choose the lenght of Δ so small that

$$\sum_{j \in c_o(t)} \frac{\partial}{\partial u} \Psi_o(\hat{u}_o(s),s)_j z_j^1 + \sum_{j \in r_o(t)} (B_o^T(s) \frac{\partial}{\partial x} \Theta_o(\hat{x}_o(s),s))_j z_j^2$$

$$\geq \gamma(|z^1| + |z^2|)/2.$$

for all (z^1,z^2) and $s \in \Delta$, where $c_o(t)$ and $r_o(t)$ correspond to $\delta = 0$. Extract from the open covering of $[0,1]$ a finite subcovering $\Delta_1,\dots,\Delta_N$ and let t_i be the centre of Δ_i.Defining

$$\delta_c = \inf \{ -\Psi_o(\hat{u}_o(s),s)_j,\ s \in \Delta_i\ ,j \notin c_o(t_i),\ i = 1,\dots,N\},$$

$$\delta_s = \inf \{ -\Theta_o(x_o(s),s)_j,\ s \in \Delta_i\ ,j \notin r_o(t_i),\ i = 1,\dots,N\},$$

the lemma holds for $0 < \delta_o < \min\{\delta_c, \delta_s\}$ and $\gamma(\delta) = \gamma/2$.Observe that $\gamma(\delta) \to 0$ as $\delta \to 0$,Q.E.D.

The next lemma is due to Malanowski [51] .

Lemma 2.6.The interval (0,1) can be split into a finite nimber of subintervals $M_k = (t_{k-1},t_k)$,$k=1,\dots,L$, to which correspond the sets of indices c^k,r^k , so that

$$\Psi_o(\hat{u}_o(t),t)_j \geq -\delta_o/2\ ,j \in c^k \quad ; \quad \Theta_o(\hat{x}_o(t),t)_j \geq -\delta_o/2\ ,j \in r^k \ ;$$

$$\Psi_o(\hat{u}_o(t),t)_j \leq -\delta_o/4\ ,j \notin c^k \quad ; \quad \Theta_o(\hat{x}_o(t),t)_j \leq -\delta_o/4\ ,j \notin r^k \ .$$

Moreover, there exists a constant $\gamma_1 > 0$ such that if $j \notin r^{k-1}$ and $j \in r^k$, then

$$\Theta_o(\hat{x}_o(t),t)_j \leqslant -\delta_o/4 \quad \text{for } t \in [t_{k-1}, t_{k-1} + \gamma_1] \ .$$

Let $m^k = \{1,\dots,q\} \setminus c^k$ and $n^k = \{1,\dots,p\} \setminus r^k$. In the sequel vectors (matrices) with superscripts $r(\cdot), n(\cdot), c(\cdot)$ and $m(\cdot)$ will denote subvectors (submatrices) corresponding to the appropriate sets of constraints.

Lemma 2.7. There exists a constant c such that for every $k = 1,\dots,L$

$$\int_{t_{k-1}}^{t_k} |\lambda_\varepsilon^{m^k}(t)|\,dt \leqslant c\varepsilon \ . \tag{30}$$

Proof. From (25) and Lemma 2.3 we conclude that

$$\hat{I}_\varepsilon \leqslant I_\varepsilon(\hat{x}_o(\cdot),\hat{u}_o(\cdot)) + \langle \lambda_\varepsilon, \Psi_o(\hat{u}_o)\rangle + c\varepsilon .$$

Combining this inequality with (26) as in the proof of Theorem 2.4 and using Lemma 2.6 we get

$$-c\varepsilon \leqslant \langle \lambda_\varepsilon, \Psi_o(\hat{u}_o)\rangle \leqslant \sum_{k=1}^{L} \int_{t_{k-1}}^{t_k} \Psi_o^{m^k}(\hat{u}_o(t),t)^T \lambda_\varepsilon^{m^k}(t)\,dt$$

$$\leqslant -\frac{\delta_o}{4} \sum_{k=1}^{L} \int_{t_{k-1}}^{t_k} \sum_{j \in m^k} \lambda_\varepsilon^{m^k}(t)_j\,dt \leqslant 0 \ .$$

This implies (30), Q.E.D.

Lemma 2.8. There exists a constant c such that for every $k = 1,\dots,L$

$$\int_{t_{k-1}}^{t_k} |\lambda_\varepsilon^{c^k}(t)|^2\,dt < c \ .$$

Proof. Using Theorem 2.4 and Lemma 2.5 we have

$$\gamma(\delta_o/2)\Big(\int_{t_{k-1}}^{t_k} |\lambda_\varepsilon^{c^k}(t)|^2 dt\Big)^{0.5} \leqslant \Big(\int_{t_{k-1}}^{t_k} \Big|\frac{\partial}{\partial u}\psi_o^{c^k}(\hat{u}_o(t),t)^T\lambda_\varepsilon^{c^k}(t)\Big|^2 dt\Big)^{0.5}$$

$$\leqslant \Big(\int_{t_{k-1}}^{t_k} \Big|\frac{\partial}{\partial u}(\psi_\varepsilon(\hat{u}_\varepsilon(t),t) - \psi_o(\hat{u}_o(t),t))\Big|^2 dt \int_{t_{k-1}}^{t_k} |\lambda_\varepsilon^{c^k}(t)|^2 dt\Big)^{0.5}$$

$$+ \Big(\int_{t_{k-1}}^{t_k} \Big|\frac{\partial}{\partial u}\psi_\varepsilon^{c^k}(\hat{u}_\varepsilon(t),t)^T\lambda_\varepsilon^{c^k}(t)\Big|^2 dt\Big)^{0.5}$$

$$\leqslant c(\sqrt{\varepsilon} + \varepsilon)\Big(\int_{t_{k-1}}^{t_k} |\lambda_\varepsilon^{c^k}(t)|^2 dt\Big)^{0.5}$$

$$+ \Big(\int_{t_{k-1}}^{t_k} \Big|\frac{\partial}{\partial u}\psi_\varepsilon^{c^k}(\hat{u}_\varepsilon(t),t)^T\lambda_\varepsilon^{c^k}(t)\Big|^2 dt\Big)^{0.5} .$$

Hence, for ε sufficiently small

$$\int_{t_{k-1}}^{t_k} |\lambda_\varepsilon^{c^k}(t)|^2 dt \leqslant c\int_{t_{k-1}}^{t_k} \Big|\frac{\partial}{\partial u}\psi_\varepsilon(\hat{u}_\varepsilon(t),t)^T\lambda_\varepsilon(t)\Big|^2 dt .$$

Taking advantage of (21) and Lemma 2.3 we complete the proof.

Introduce the notation: $\Delta x = \hat{x}_\varepsilon - \hat{x}_o$, $\Delta u = \hat{u}_\varepsilon - \hat{u}_o$, $\Delta q = q_\varepsilon - q_o$, $\Delta\lambda = \lambda_\varepsilon - \lambda_o$, $\Delta\nu = \nu_\varepsilon - \nu_o$, $\Delta p = p_\varepsilon - p_o$.

__Lemma 2.9.__ The following estimation holds

$$\|\Delta q\|_C + \|\Delta\nu\|_{L^1} + \|\Delta\lambda\|_{L^1} + \|\Delta p\|_{L^1} \leqslant c(\|\Delta u\| + \varepsilon). \qquad (31)$$

Proof. Since $\lambda_o^{c(t)}(t) = 0$, from (21) we obtain that for every $t \in [0,1]$

$$\Big[-\frac{\partial}{\partial u}\psi_o^{c(t)}(\hat{u}_o(t),t)^T,\ B_o^T(t)\ \frac{\partial}{\partial x}\Theta_o^{r(t)}(\hat{x}_o(t),t)^T\Big]\begin{bmatrix}\Delta\lambda^{c(t)}(t)\\ \\ \Delta\nu^{r(t)}(t)\end{bmatrix}$$

$$= \frac{\partial}{\partial u}(f_\varepsilon(\hat{x}_\varepsilon(t),\hat{u}_\varepsilon(t),t) - f_o(\hat{x}_o(t),\hat{u}_o(t),t)) + B_o^T(t)\Delta q(t)$$

$$+ (B_\varepsilon(t) - B_o(t))^T q_\varepsilon(t) + (\frac{\partial}{\partial u}(\Psi_\varepsilon(\hat{u}_\varepsilon(t),t)$$

$$- \Psi_o(\hat{u}_o(t),t))^{c(t)})^T \lambda_\varepsilon^{c(t)}(t)$$

$$+ (\frac{\partial}{\partial u}\Psi_\varepsilon(\hat{u}_\varepsilon(t),t)^{m(t)})^T \lambda_\varepsilon^{m(t)}(t)$$

$$- (B_\varepsilon^T(t)\frac{\partial}{\partial x}\Theta_\varepsilon^T(\hat{x}_\varepsilon(t),t) - B_o^T(t)\frac{\partial}{\partial x}\Theta_o^T(\hat{x}_o(t),t))\,\nu_\varepsilon(t)$$

$$- B_o^T(t)\frac{\partial}{\partial x}\Theta_o^T(\hat{x}_o(t),t)^{n(t)}\Delta\nu^{n(t)}(t) .$$

Applying lemmas 2.3 and 2.5 we conclude that

$$|\Delta\lambda^{c(t)}(t)| + |\Delta\nu^{r(t)}(t)| \leqslant c(|\Delta u(t)| + |\Delta x(t)| + |\Delta q(t)| + |\lambda_\varepsilon^{m(t)}(t)| + |\Delta u(t)|\,|\lambda_\varepsilon^{c(t)}(t)| + |\Delta\nu^{n(t)}(t)| + \varepsilon) . \qquad (32)$$

Since $\nu_\varepsilon(1) - \nu_o(1) = 0$, from (23) we get $\Delta\nu^{n^L}(t) = 0$ for $t \in M_L$. Hence, from lemmas 2.3 and 2.7

$$\int_{t_{L-1}}^{t_L} (|\Delta\lambda(t)| + |\Delta\nu(t)|)dt \leqslant c(\int_{t_{L-1}}^{t_L} (|\Delta u(t)| + |\Delta x(t)| + |\Delta q(t)| + |\Delta u(t)|\,|\lambda_\varepsilon^{c(t)}(t)|)dt + \varepsilon) . (33)$$

Choose

$$\gamma_2 \in (0, \min\{\gamma_1, \min(\text{meas } M_k), k=1,\dots,L\}).$$

From Theorem 2.4 and lemmas 2.1 and 2.6 it follows that

$$\Theta_\varepsilon(\hat{x}_\varepsilon(t),t)_j < 0 \quad \text{for } t \in [t_{k-1}, t_{k-1} + \gamma_2]$$

when $j \notin r^{k-1}$ and $j \in r^{k}$. Then the complementary slackness condition (23) yields

$$\Delta\nu^{n(t)}(t) = \Delta\nu^{n(t)}(s) \quad \text{for } t \in M_{k-1} \text{ and } s \in [t_{k-1}, t_{k-1} + \gamma_2]. \qquad (34)$$

If

$$h = \gamma_2 / \max_k(\text{ meas } M_k) ,$$

then

$$t_k + h(t - t_{k-1}) \leqslant t_{k+1} \quad \text{for } t \in M_k .$$

Let us assume that $t \in M_{L-1}$. Taking advantage of (32) and (34) one gets

$$\begin{aligned} |\Delta\lambda^{c(t)}(t)| + |\Delta\nu(t)| &\leqslant c(|\Delta u(t)| + |\Delta x(t)| + |\Delta q(t)| + |\lambda_\varepsilon^{m^{L-1}}(t)| \\ &\quad + |\Delta u(t)||\lambda_\varepsilon^{c^{L-1}}(t)| + |\Delta\nu^{n^{L-1}}(t)| + \varepsilon) \\ &\leqslant c(|\Delta u(t)| + |\Delta x(t)| + |\Delta q(t)| \\ &\quad + |\lambda_\varepsilon^{m^{L-1}}(t)| + |\Delta u(t)||\lambda_\varepsilon^{c^{L-1}}(t)| \\ &\quad + |\Delta\nu(t_{L-1} + h(t - t_{L-2}))| + \varepsilon) . \end{aligned} \qquad (35)$$

Using the inequality

$$\int_{t_{L-2}}^{t_{L-1}} |\Delta\nu(t_{L-1} + h(t - t_{L-2}))|\, dt \leqslant \int_{t_{L-1}}^{t_L} |\Delta\nu(t)|\, dt ,$$

integrating the both sides of (35) in $[t_{L-2}, t_{L-1}]$,taking advantage of Lemma 2.7 and substituting (33) in (35) we obtain

$$\int_{t_{L-2}}^{t_{L-1}} (|\Delta\lambda(t)| + |\Delta\nu(t)|)dt \leqslant c(\int_{t_{L-2}}^{t_L} (|\Delta u(t)| + |\Delta u(t)|\, |\lambda_\varepsilon^{c(t)}(t)|$$

$$+ |\Delta x(t)| + |\Delta q(t)|)dt + \varepsilon) .$$

Procceding in the same manner by induction we get

$$\int_{t_{k-1}}^{t_k}(|\Delta\lambda(t)| + |\Delta\nu(t)|)dt \leqslant c(\int_{t_{k-1}}^{1}(|\Delta u(t)| + |\Delta u(t)||\lambda_\varepsilon^{c(t)}(t)|$$

$$+|\Delta x(t)| + |\Delta q(t)|)dt + \varepsilon). \qquad (36)$$

Similarly, if $t' \in M_k$

$$\int_{t'}^{t_k}(|\Delta\lambda(t)| + |\Delta\nu(t)|)dt \leqslant c(\int_{t'}^{1}(|\Delta u(t)| + |\Delta u(t)||\lambda_\varepsilon^{c(t)}(t)|$$

$$+|\Delta x(t)| + |\Delta q(t)|)dt + \varepsilon). \qquad (37)$$

From (22) we have

$$|\Delta q(t)| \leqslant c(\int_t^1(|\Delta q(s)| + |\Delta u(s)| + |\Delta x(s)| + |\Delta\nu(s)|)ds + \varepsilon).$$

Taking advantage of (37) and using Gronwall lemma and lemmas 2.1 and 2.8 we obtain (31), Q.E.D.

Thus, using Lemma 2.4 and substituting (21) in (29) we complete the proof of Theorem 2.6.

Moreover, from lemmas 2.1 and 2.9 we get

Corollary 2.1

$$\|\hat{x}_\varepsilon - \hat{x}_0\|_C + \|q_\varepsilon - \hat{q}_0\|_C + \|p_\varepsilon - \hat{p}_0\|_{L^1}$$

$$+ \|\nu_\varepsilon - \nu_0\|_{L^1} + \|\lambda_\varepsilon - \lambda_0\|_{L^1} \leqslant c\varepsilon .$$

Obviously, the only possible refinement of this result is to proof Lipschitz continuity of the optimal control in a metric stronger that L^2 and, respectively, Lipschitz continuity of the dual variables in a metric stronger than L^1. This problem is investigated in the next paragraph.

Theorem 2.6 was announced in the author's paper [16].

2.3.4. Control constraints

We take up the problem (CP_ε) in a somewhat different form assuming that the state constraints are vacuous.For simplicity and ease of notation we suppose that the performance index, the control mattix and the control constraints do not contain a perturbation parameter.

We seek a solution of the following problem: minimize the functional

$$I(x(\cdot),u(\cdot)) = g(x(1)) + \int_0^1 f(x(t),u(t),t)dt , \qquad (38)$$

subject to the constraints

$$\dot{x} = A_\varepsilon(t)x + B(t)u , \ t\in[0,1] , \quad x(0) = x^0 , \qquad (39)$$

$$u(\cdot)\in U = \{u(\cdot)\in L^2(R^m),\ u(t)\in V(t) \text{ for a.e. } t\in[0,1]\}, \qquad (40)$$

$$x(\cdot)\in H^1(R^n) ,$$

where $\varepsilon\in[0, \varepsilon_0]$,assuming that:

A8. The matrices $A_\varepsilon(t)$ and $B(t)$ are continuous; $A_\varepsilon(t)$ is Lipschitz continuous with respect to ε at $\varepsilon = 0$ with a Lipschitz constant in $L^1(R^1)$.

A9(i). The function $g(\cdot)$ is convex and differentiable;the derivative $g'(\cdot)$ is Lipschitz continuous.The function $f(\cdot)$ is continuous, it is differenriable with respect to (x,u) and the derivative $\frac{\partial}{\partial(x,u)}f(\cdot)$ is continuous; $\frac{\partial}{\partial(x,u)}f(\cdot)$ is Lipschitz continuous with respect to (x,u) with a Lipschitz constant in $L^2(R^1)$.There exists a constant $\varkappa>0$ such that for every $z = (x,u)$ and $w = (y,v)\in R^{n+m}$ and for every $\alpha\in[0,1]$ and $t\in[0,1]$

$$f(\alpha z + (1-\alpha)w,t) \leqslant \alpha f(z,t) + (1-\alpha)f(w,t) - \alpha(1-\alpha)\varkappa|u - v|^2 .$$

(ii) The set $V(t)$ is closed and convex for every $t\in[0,1]$; there exists a function $\bar{u}(\cdot)\in L^\infty(R^m)$ such that $\bar{u}(t)\in V(t)$ for a.e.$t\in[0,1]$.

Let $J_\varepsilon(u(\cdot)) = I(x^u_\varepsilon(\cdot),u(\cdot))$, where $x^u_\varepsilon(\cdot)$ is determined by (39) for given $u(\cdot)$ and ε .Since the functional $J_\varepsilon(\cdot)$ is strongly convex in $L^2(R^m)$ and the set U is closed and convex, there exists unique solution $(\hat{x}_\varepsilon(\cdot),\hat{u}_\varepsilon(\cdot))$ for every $\varepsilon\in[0,\varepsilon_0]$ .We assume in addition that the optimal control $\hat{u}_\varepsilon(\cdot)$ satisfies the Maximum principle in a normal form,that is

A10. For almost all $t\in[0,1]$

$$H_\varepsilon(\hat{u}_\varepsilon(t),t) = \max_{u\in V(t)} H_\varepsilon(u,t) ,$$

where

$$H_\varepsilon(u,t) = -f(\hat{x}_\varepsilon(t),u,t) + p^T_\varepsilon(t)B(t)u ,$$

and $p_\varepsilon(\cdot)$ solves the adjoint equation

$$\dot{p} = -A^T_\varepsilon(t)p + \frac{\partial}{\partial x}f(x(t),u(t),t) ,$$

$$p(1) = -g'(x(1)) \qquad (41)$$

for $(x(t),u(t)) = (\hat{x}_\varepsilon(t),\hat{u}_\varepsilon(t))$.

Let us recall the following standard result,see for instance Vasilev [71] ,p.258 :

<u>Lemma 2.10</u>. The functional $J_\varepsilon(\cdot)$ is Frechet differentiable in $L^2(R^m)$ and the derivative $J'_\varepsilon(\cdot)$ is given by

$$(J'_\varepsilon(u(\cdot)))(t) = \frac{\partial}{\partial u}f(x_\varepsilon(t),u(t),t) - B^T(t)p_\varepsilon(t) , \qquad (42)$$

where $p_\varepsilon(\cdot)$ is the solution of (41) and $x_\varepsilon(\cdot)$ corresponds to $u(\cdot)$ and ε .

<u>Theorem 2.7</u>. The following estimation holds

$$\|\hat{u}_\varepsilon - \hat{u}_0\|_{L^\infty} \leq c\varepsilon . \qquad (43)$$

Proof.Obviously, the statements of lemmas 2.1 and 2.2 hold.Moreover, by (41) we deduce that $\limsup_{\varepsilon\to 0} \|p_\varepsilon\|_C < +\infty$ .The strong concavity of the Hamiltonian yields

$$2\varkappa|\hat{u}_\varepsilon(t) - \bar{u}(t)| \leq \left|\frac{\partial}{\partial u} H_\varepsilon(\bar{u}(t),t)\right| \qquad (44)$$

Then, for every $\varepsilon\in[0,\varepsilon_0]$ the optimal control $\hat{u}_\varepsilon(\cdot)\in L^\infty(R^m)$.

The problem (38)-(40) can be thought of as a problem with perturbed functional.Therefore, one can directly apply the estimations from Chapter 1.

Let $x_\varepsilon(\cdot)$ be determined by (39) for $u(\cdot) = \hat{u}_0(\cdot)$.The Gronwall lemma yields

$$\|x_\varepsilon - \hat{x}_0\|_C = O(\varepsilon) \quad . \tag{45}$$

Let $\tilde{p}_\varepsilon(\cdot)$ be the solution of (41) corresponding to $(x_\varepsilon(t),\hat{u}_0(t))$. Applying again the Gronwall lemma we get

$$\|\tilde{p}_\varepsilon - p_0\|_C = O(\varepsilon) \quad . \tag{46}$$

Hence

$$\|J'_\varepsilon(\hat{u}_0(\cdot)) - J'_0(\hat{u}_0(\cdot))\| = O(\varepsilon) \ .$$

Thus,by Corollary 1.3

$$\|\hat{u}_\varepsilon - \hat{u}_0\| = O(\varepsilon) \ .$$

From Lemma 2.1

$$\|\hat{x}_\varepsilon - \hat{x}_0\|_C = O(\varepsilon) \ . \tag{47}$$

Similarly

$$\|p_\varepsilon - p_0\|_C = O(\varepsilon) \ . \tag{48}$$

One can apply Corollary 1.3 to the Maximum principle,that is

$$2\varkappa|\hat{u}_\varepsilon(t) - \hat{u}_0(t)| \leqslant \left|\frac{\partial}{\partial u}(H_\varepsilon(\hat{u}_0(t),t) - H_0(\hat{u}_0(t),t))\right|$$

for almost all $t\in[0,1]$ .Thus,substituting (47) and (48) in the above inequality we obtain (43),Q.E.D.

<u>Remark 2.5</u>. Since

$$\limsup_{\varepsilon\to 0} \left(\|\hat{x}_\varepsilon\|_C + \|\hat{u}_\varepsilon\|_{L^\infty}\right) < +\infty \ ,$$

the Lipschitz continuity of the derivatives of $g(\cdot)$ and $f(\cdot)$ can be replaced by a local Lipschitz continuity.

Now we slightly change the problem considered strengthening the above result to the uniform metric and estimating the Lipschitz constants.In order to avoid long expression, instead of (38) we take the following quadratic functional

$$I(x(\cdot),u(\cdot)) = 0.5\int_0^1 (|x(t)|^2 + |u(t)|^2)dt ,$$

assuming that the matrix $A_\varepsilon(t)$ is Lipschitz continuous with respect to ε at $\varepsilon = 0$ with a Lipschitz constant L_a independent of t.The constraining set is not time-varying but perturbed,that is

$$U_\varepsilon = \{u(\cdot) \in L^2(R^m),\ u(t) \in V_\varepsilon \quad \text{for a.e. } t \in [0,1]\},$$

where:

A11. For every $\varepsilon \in [0, \varepsilon_0]$ the set V_ε is closed and convex and contains the origin of R^m.

It is classically known that in this case the Maximum principle holds with Hamiltonian

$$H_\varepsilon(u,t) = -0,5\,|u|^2 + p_\varepsilon^T(t)B(t)u .$$

Define

$$d(\varepsilon) = \min_{u \in V_\varepsilon} \max_{t \in [0,1]} |u-\hat{u}_0(t)| + \min_{u \in V_0} \max_{t \in [0,1]} |u-\hat{u}_\varepsilon(t)| .$$

One can easily prove that

$$\min_{u \in U_0} \|u-\hat{u}_\varepsilon\| + \min_{u \in U_\varepsilon} \|u-\hat{u}_0\| \leqslant d(\varepsilon) .$$

<u>Theorem 2.8</u>. Let $a \geqslant \sup \|A_\varepsilon\|_C$ for $\varepsilon \in |0,\varepsilon_0|$ and

$$c_1 \geqslant e^a(L_a\|p_0\|_{L^1} + e^a(L_a\|\hat{x}_0\|_{L^1} + \|B\|^2 e^a(L_a\|p_0\|_{L^1} + e^a L_a\|\hat{x}_0\|_{L^1}))),$$

$$c_2 \geqslant (2e^{2a}(1 + \|B\|^2 e^{2a})\,|x^0|\,)^{1/2}(e^{2a}\|B\|^{3/2} + \|B\|_C\ ) .$$

Then for every $\varepsilon \in [0, \varepsilon_0]$

$$\|\hat{u}_\varepsilon - \hat{u}_0\|_C \leqslant c_1 \varepsilon + c_2 d(\varepsilon)^{1/2} .$$

Proof. From (44) with $\bar{u}(\cdot) = 0$ it follows that for every $\varepsilon \in [0, \varepsilon_0]$ the optimal control $\hat{u}_\varepsilon(\cdot)$ is uniformly bounded on a set $E \subset (0,1)$ with measure one.Let $t_k \in E$ for $k=1, \dots$,and let $\lim_{k\to\infty} t_k = t' \in [0,1] \setminus E$. Clearly, there exists a subsequence t_k' and a limit $u' \in V_\varepsilon$ such that

$$H_\varepsilon(u', t') = \max_{u \in V_\varepsilon} H_\varepsilon(u, t') .$$

Defining $\hat{u}_\varepsilon(t') = u'$we conclude that there exists a function which satisfies the Maximum principle for all $t \in [0,1]$ and which coincides with the optimal control for almost all $t \in [0,1]$ .Obviously,this function is an optimal control.Moreover, from Corollary 1.3 we get

$$|\hat{u}_\varepsilon(t_1) - \hat{u}_\varepsilon(t_2)| \leqslant |\frac{\partial}{\partial u}(H_\varepsilon(\hat{u}_\varepsilon(t_2), t_2) - H_\varepsilon(\hat{u}_\varepsilon(t_2), t_1))|$$

for all $t_1, t_2 \in [0,1]$ and for every fixed ε .Hence,there exists an optimal control which is a continuous function on $[0,1]$.

By repeating the arguments of the proof of Theorem 2.7 and estimating the constants in (45) and (46) we obtain

$$\|J'_\varepsilon(\hat{u}_0) - J'_0(\hat{u}_0)\| \leqslant c_3 \varepsilon ,$$

where

$$c_3 = \|B\| e^a (e^a L_a \|\hat{x}_0\|_{L^1} + L_a \|p_0\|_{L^1}) .$$

Taking advantage of Proposition 1.2 in Chapter 1 we get

$$\|\hat{u}_\varepsilon - \hat{u}_0\| \leqslant c_3 \varepsilon + c_4 d(\varepsilon)^{1/2} , \tag{49}$$

where c_4 fulfilles

$$c_4 = (\sup \|\frac{\partial}{\partial u} H_\varepsilon(\hat{u}_\varepsilon)\|, \varepsilon \in [0, \varepsilon_0])^{1/2} .$$

Substituting $\bar{u}(t) \equiv 0$ in (17) and (18) we obtain

$$\|\hat{u}_\varepsilon\| \leqslant e^{2a}\|B\|\,|x^o| \ .$$

Since

$$\|\hat{x}_\varepsilon\|_C \leqslant e^a(|x^o| + \|B\|\|\hat{u}_\varepsilon\|)$$

we get

$$\|p_\varepsilon\|_C \leqslant e^{2a}(1 + \|B\|^2 e^{2a})|x^o| \ .$$

Using this relation and

$$\|\hat{u}_\varepsilon\| \leqslant \|B\|\|p_\varepsilon\|$$

one can find an expression for c_4 .

Applying again Proposition 1.2 to the Maximum principle we have

$$\begin{aligned}|\hat{u}_\varepsilon(t) - \hat{u}_o(t)| \leqslant |B^T(t)(p_\varepsilon(t) - p_o(t))| \\ + (|\frac{\partial}{\partial u}H_\varepsilon(\hat{u}_\varepsilon(t),t)||v_\varepsilon - \hat{u}_o(t)| \\ + (|\frac{\partial}{\partial u}H_o(\hat{u}_o(t),t)||v_o - \hat{u}_\varepsilon(t)|)^{0.5}, \qquad (50)\end{aligned}$$

where $v_\varepsilon \in V_\varepsilon$ and $v_o \in V_o$ are arbitrarily chosen.From (49),using twice the Gronwall lemma,one can obtain an estimate for $\|p_\varepsilon - p_o\|_C$. This estimate,applied to (50) gives us the desired result.

Remark 2.6.The proofs of the Lipschitz continuity in the last two paragraphs provide a basis for estimating the Lipschitz constant,as it was done in Theorem 2.8.For that purpose it is necessary to repeat every step of the proofs making appropriate estimations. For such general problems,however, the obtained constants may be rough.It is more relevant to use the metod for determining the Lipschitz constants in every concrete case.

2.3.5. A fixed final state problem.

In the previous paragraph we applied directly the scheme from Chapter 1 to optimal control problems with local constraints.Here we demonstrate an application of this scheme to the following problem with fixed initial and final state :

$$I(x(\cdot),u(\cdot)) = 0.5 \int_0^1 (|x(t)|^2 + |u(t)|^2)dt \longrightarrow \inf$$

$$\dot{x} = A_\varepsilon(t)x + B(t)u \ , x(0) = x^o \ , \ u(\cdot) \in L^2(R^m) \ , \quad (51)$$

$$x(1) = x^1 \ , \quad (52)$$

where the matrices $A_\varepsilon(t)$ and $B(t)$ satisfy A8. In addition we assume that

A12. The system (51) with $\varepsilon = 0$ is controllable on $[0,1]$,that is for every $x^1 \in R^n$ there exists a control $u(\cdot) \in L^2(R^m)$ driving the state from x^o to x^1.

It is classicaly known that the controllability is equivalent to the nonsingularity of the matrix

$$M_\varepsilon = \int_0^1 E_\varepsilon(1,t)B(t)B^T(t)E_\varepsilon^T(1,t)dt \ ,$$

where $E_\varepsilon(t,s)$ is the fundamental matrix solution of the homogeneus equation $\dot{x} = A_\varepsilon(t)x$.Since

$$\| E_\varepsilon(1,\cdot) - E_o(1,\cdot)\|_C = O(\varepsilon) \quad (53)$$

the system (51) is controllable for sufficiently small ε .This means that for small ε the optimal control $\hat{u}_\varepsilon(\cdot)$ exists and is unique. Moreover, by repeating the proof of Theorem 2.8 ,the optimal control can be thought of as a continuous function of the time.

Let us define the operator $F_\varepsilon : L^2(R^m) \longrightarrow R^n$ as

$$F_\varepsilon u(\cdot) = \int_0^1 E_\varepsilon(1,t)B(t)u(t)dt \ .$$

Then the constraint (52) can be considered as an integral equality constraint for the control,that is

$$F_\varepsilon u(\cdot) = x^1 - E_\varepsilon(1,0)x^o. \quad (54)$$

Hence, we can apply the results from Section 1.3.The optimality condition has the form

$$\frac{\partial}{\partial u} H_\varepsilon(\hat{u}_\varepsilon(t),t) + B^T(t)E_\varepsilon^T(1,t)\mu^\varepsilon = 0 ,$$

where the Lagrange multiplier μ^ε is associated with the final state constraint (54).

Theorem 2.9.

$$|\hat{u}_\varepsilon - \hat{u}_o|_C \leqslant c\varepsilon .$$

Proof.Using (53) we get

$$\| F_\varepsilon - F_o \| = O(\varepsilon),$$

where $\| \cdot \|$ denotes the operator norm.Let e_k be a bounded sequence in R^n and let $\lim_{k\to\infty} \varepsilon_k = 0$.For large k define the sequence of controls

$$\bar{u}_k(t) = B^T(t)E_{\varepsilon_k}^T(1,t)M_{\varepsilon_k}^{-1}(e_k - E_{\varepsilon_k}(1,0)x^o) .$$

Then the sequence $\bar{u}_k(\cdot)$ is bounded in $L^2(R^m)$ and

$$F_{\varepsilon_k}\bar{u}_k(\cdot) = e_k .$$

Hence, condition A3 in Section 1.3 holds.Observe that the condition A4 in Section 1.3 is equivalent to the controllability assumption. Using Lemma 2.10, (53) and the Gronwall lemma to the state and the adjoint equations we conclude that there exists a constant K such that

$$\| J_\varepsilon'(\hat{u}_\varepsilon(\cdot)) - J_\varepsilon'(\hat{u}_o(\cdot))\| \leqslant K\|\hat{u}_\varepsilon - \hat{u}_o\| .$$

Therefore, one can apply Proposition 1.4 obtaining

$$|\hat{u}_\varepsilon - \hat{u}_o\| + |\mu^\varepsilon - \mu^o| = O(\varepsilon) .$$

The further proof is analogous to the second part of the proof of Theorem 2.7.

CHAPTER 3

SINGULAR PERTURBATIONS

3.1. Introduction

A complex system may consist of many interacting subprocesses with widely different time scales. The different dynamics leads to the effect that by the time the slow processes change noticeably, the fast processes have already reached their steady-states.

If we consider a dynamic system modelled by a differential equation, such a situation can be described mathematically by a small parameter in the derivatives of the fast variables. This parameter represents a perturbation which may change essentially the structure of the model. At the limit, when the parameter becomes zero, the differential equation associated with the fast subprocess reduces to algebraic one, that is the order of the system changes. This is in fact a change of the state space. Such a change is often accompanied by pathological mathematical effects as lack of definiteness, boundary layers etc.

The perturbation provided by a small parameter in the derivatives of a part of the states is commonly called singular perturbation. As examples of real-life processes described by singularly perturbed differential equations, see the voltage regulator model and the transformer model presented in Kokotovic et al. [45], and the model of a spinning shell given in Moiseev [54].

This chapter is concerned with singularly perturbed control systems described by linear ordinary differential equations

$$\dot{x} = A_1(t)x + A_2(t)y + B_1(t)u \ , \quad x(0) = x^0 \ , \tag{1a}$$

$$\beta\dot{y} = A_3(t)x + A_4(t)y + B_2(t)u \ , \quad y(0) = y^0 \ , \tag{1b}$$

where $x(t) \in R^m$, $y(t) \in R^n$, $u(t) \in R^r$, the time $t \in [0,1]$, and β is a positive "small" parameter.The states $x(t)$ and $y(t)$ represent the slow and the fast phenomena, respectively.

For $\beta = 0$ the order $m+n$ of the system (1) reduces to m, that is (1) becomes

$$\dot{x} = A_1(t)x + A_2(t)y + B_1(t)u \ , \ x(0) = x^0 \ , \tag{2a}$$

$$0 = A_3(t)x + A_4(t)y + B_2(t)u \ , \tag{2b}$$

for $t \in [0,1]$, where the initial condition $y(0) = y^0$ is dropped since there is no freedom to satisfy it.Assuming that the matrix $A_4(t)$ is nonsingular and solving the equation (2b) with respect to y we obtain the <u>reduced model</u>

$$\dot{x} = A_0(t)x + B_0(t)u \ , \ x(0) = x^0 \ , \tag{3}$$

where $A_0 = A_1 - A_2A_4^{-1}A_3$ and $B_0 = B_1 - A_2A_4^{-1}B_2$.

Clearly, the order reduction may lead to an essential simplification of the original model.On the other hand, differential equations containing small parameters in the derivative, commonly called stiff equations, often need sophisticated numerical technique.Thus, it is important to know when the low order model (3) provides a "good" approximation to the basic model (1).

Recently, the order reduction in optimal control has received a great deal of attention; the reader can find comprehensive surveys in Kokotovič et al. [44] and in Vasileva and Dmitriev [73] Here we present author's results, a part of which are published in [15] and [17] , and results by the author and V.Veliov [19] , [20] .We refer to related papers throughout the exposition.

An outline of the chapter follows:

Section 3.2 - Presents three auxiliary lemmas.

Section 3.3 - Considers the convergence of the reachable sets of singularly perturbed control systems with restricted and free controls.

Section 3.4 - Deals with the performance and solution well-posedness of the optimal control problems of Mayer and Lagrange.

Section 3.5 - Gives estimations for the optimal value and for the optimal control.

Throughout the chapter we shall use the notation of the previous chapter.All the constants which are not dependent of the time t and the perturbation parameter β will be denoted by c_i, $i = 0,1,\dots,$ or simply by c.

3.2. Preliminary lemmas

We first start from some auxiliary results.Let $\varphi_o(\cdot) \in L^2(R^m)$, $\psi_o(\cdot) \in L^2(R^n)$ and β_k, $\lim_{k\to\infty} \beta_k = 0, \Delta\varphi_k(\cdot), \Delta\psi_k(\cdot)$, v_k and w_k be given sequences.Denote by $(x_k(\cdot), y_k(\cdot))$ the solution of the equation

$$\dot{x} = A_1(t)x + A_2(t)y + \varphi_o(t) + \Delta\varphi_k(t),\ x(0) = v_k\ ,$$
$$\beta_k\dot{y} = A_3(t)x + A_4(t)y + \psi_o(t) + \Delta\psi_k(t),\ y(0) = w_k\ , \qquad (4)$$

and by $(x_o(\cdot), y_o(\cdot))$ the solution of

$$\dot{x} = A_1(t)x + A_2(t)y + \varphi_o(t)\ ,\ x(0) = v_o\ ,$$
$$0 = A_3(t)x + A_4(t)y + \psi_o(t)\ , \qquad (5)$$

for $t \in [0,1]$ .We assume that:

A1. The matrices $A_i(t)$ are continuous on $[0,1]$ .All the eigenvalues of the matrix $A_4(t)$ have negative real parts for every $t \in [0,1]$.

Lemma 3.1.

(i) Let $\lim_{k\to\infty} v_k = v_o$, $w_k = w_k'/\beta_k$, $\lim_{k\to\infty} w_k' = w_o'$ and the sequences $\Delta\varphi_k(\cdot), \Delta\psi_k(\cdot)$ are weakly convergent to zero in $L^2(R^m)$ and $L^2(R^n)$ respectively.Let $x_o(\cdot)$ solve (5) with the initial condition

$$x(0) = v_o - A_2(0)A_4^{-1}(0)w_o' \ .$$

Then the sequence $x_k(\cdot)$ is uniformly bounded in $[0,1]$ and for every $\Theta \in (0,1)$

$$\lim_{k\to\infty} \max_{\Theta \leq t \leq 1} |x_k(t) - x_o(t)| = 0. \qquad (6)$$

(ii) If, additionally, $w_o' = 0$, then

$$\lim_{k\to\infty} \|x_k - x_o\|_C = 0. \qquad (7)$$

(iii) Let $\lim_{k\to\infty} \beta_k\sqrt{w_k} = 0$ and all the above conditions hold. Then the sequence $y_k(\cdot)$ is L^2-weakly convergent to $y_o(\cdot)$.

(iv) Let $\lim_{k\to\infty} v_k = v_o$ and $\lim_{k\to\infty} \beta_k\sqrt{w_k} = 0$. Then there exists a constant c such that for k sufficiently large

$$\|x_k - x_o\|_C + \|y_k - y_o\| \leqslant c(\|\Delta\varphi_k\| + \|\Delta\psi_k\| + \delta_k), \qquad (8)$$

where $\lim_{k\to\infty} \delta_k = 0$.

(v) Let the above conditions be satisfied and, additionally: $\psi_o(\cdot) \in C(R^n)$, the sequence w_k be bounded, and for every $\theta_1 \in (0,1)$

$$\lim_{k\to\infty} \max_{0\leqslant t\leqslant \theta_1} |\Delta\psi_k(t)| = 0.$$

Then for every $\theta \in (0,\theta_1/2)$

$$\lim_{k\to\infty} \max_{\theta\leqslant t\leqslant 1-\theta} |y_k(t) - y_o(t)| = 0. \qquad (9)$$

Proof. Let $Y(t,s,\beta_k)$ be the fundamental matrix solution of the equation $\beta_k\dot{y} = A_4(t)y$, normalized at $t = s$. By Vasileva and Butuzov [72], there exist constants $\sigma_o, \sigma > 0$ such that

$$|Y(t,s,\beta_k)| \leqslant \sigma_o \exp\left(-\sigma\frac{t-s}{\beta_k}\right) \qquad (10)$$

for all $t,s \in [0,1]$, $t \geqslant s$. Denoting $\Delta x_k = x_k - x_o$, $\Delta y_k = y_k - y_o$, we have

$$\Delta x_k(t) = v_k - v_o + A_2(0)A_4^{-1}(0)w_o'$$

$$+ \int_0^t (A_1(s)\Delta x_k(s) + A_2(s)\Delta y_k(s) + \Delta\varphi_k(s))ds, \qquad (11)$$

$$y_k(t) = Y(t,0,\beta_k)w_k + \frac{1}{\beta_k}\int_0^t Y(t,s,\beta_k)(A_3(s)\Delta x_k(s)$$

$$+\Delta\psi_k(s) - A_4(s)y_o(s))ds - y_o(t). \qquad (12)$$

In the sequel we shall use the following standard result:if $p(\cdot) \in L^1(R^1)$, $q(\cdot) \in L^2(R^1)$ and

$$r(t) = \int_0^t p(t - s)q(s)ds ,$$

then

$$\|r\| \leq \|p\|_{L^1}\|q\| . \tag{13}$$

Let δ be an arbitrary positive number.Choose a function $y^\delta(\cdot)$ from $C^1(R^n)$ such that $\|y^\delta - y_o\| < \delta$ .In view of (10), for every $t \in [0,1]$

$$\left|\int_0^t \frac{\partial}{\partial s} Y(t,s,\beta_k)y^\delta(s)ds - y_o(t)\right|$$

$$\leq c(|y^\delta(t) - y_o(t)| + |y^\delta(0)| \exp(-\sigma \frac{t}{\beta_k}) + \beta_k \|\dot{y}^\delta\|_C) . \tag{14}$$

Let

$$\bar{y}_k(t) = \frac{1}{\beta_k} \int_0^t Y(t,s,\beta_k)A_4(s)y_o(s)ds$$

Since δ can be arbitrarily small, from (13) and (14), integrating by parts we obtain

$$\lim_{k \to \infty} \|\bar{y}_k - y_o\| = 0. \tag{15}$$

For an arbitrary but fixed $\varepsilon > 0$ one can find matrices $A_2^\varepsilon(t)$ and $A^\varepsilon(t)$,which elements are smooth in the time, such that

$$\|A_2 - A_2^\varepsilon\|_C < \varepsilon \quad \text{and} \quad \|A_4^{-1} - A^\varepsilon\|_C < \varepsilon .$$

From

$$\frac{1}{\beta_k} \int_0^t A_2(s)Y(s,0,\beta_k)ds = \int_0^t A_2(s)A_4^{-1}(s) \frac{\partial}{\partial s} Y(s,0,\beta_k)ds$$

$$= \frac{1}{\beta_k} \int_0^t (A_2(s)A_4^{-1}(s) - A_2^\varepsilon(s)A^\varepsilon(s))A_4(s)Y(s,0,\beta_k)ds$$

$$+ \int_0^t A_2^\varepsilon(s)\; A^\varepsilon(s)\; \frac{\partial}{\partial s}Y(s,0,\beta_k)ds \;,$$

integrating by parts and taking advantage of (10) we get

$$|\frac{1}{\beta_k} \int_0^t A_2(s)Y(s,0,\beta_k)w_k' ds + A_2(0)A_4^{-1}(0)w_o'|$$

$$\leqslant c(\varepsilon + |w_k'| \exp(-\sigma \frac{t}{\beta_k}) + c_1(\varepsilon)\beta_k + |w_k' - w_o'|). \tag{16}$$

where $c_1(\varepsilon) \geqslant \| \frac{d}{dt}(A_2^\varepsilon A^\varepsilon)\|_C$. Denote

$$\xi_k(t) = A_3(t)\Delta x_k(t) + \Delta\Psi_k(t)$$

and

$$\eta_k(t) = \frac{1}{\beta_k} \int_0^t Y(t,s,\beta_k)\xi_k(s)ds \;.$$

Applying (13) we get

$$\|\eta_k\| \leqslant \frac{\sigma_0}{\sigma} \|\xi_k\| \;. \tag{17}$$

Using (10),(13),(17),Hoelder inequality and integrating by parts we obtain

$$|\int_0^t A_2(s)\eta_k(s)ds| \leqslant \int_0^t |A_2(s) - A_2^\varepsilon(s)||\eta_k(s)|ds$$

$$+ |\int_0^t A_2^\varepsilon(s)A^\varepsilon(s)A_4(s)\eta_k(s)ds|$$

$$+ \int_0^t |A_2^\varepsilon(s)||A_4^{-1}(s) - A^\varepsilon(s)||A_4(s)\eta_k(s)|ds$$

$$\leqslant c\varepsilon\|\xi_k\| + |\int_0^t A_2^\varepsilon(s)A^\varepsilon(s) \frac{d}{ds} \int_0^s Y(s,v,\beta_k)\xi_k(v)dvds|$$

$$+\left|\int_0^t A_2^\varepsilon(s)A^\varepsilon(s)\xi_k(s)ds\right|$$

$$\leqslant c\varepsilon\|\xi_k\| + \beta_k\left(|A_2^\varepsilon(t)A^\varepsilon(t)\eta_k(t)| + \left|\int_0^t \frac{d}{ds}(A_2^\varepsilon(s)A^\varepsilon(s))\eta_k(s)ds\right|\right)$$

$$+\left|\int_0^t A_2^\varepsilon(s)A^\varepsilon(s)\xi_k(s)ds\right|$$

$$\leqslant c\left((\varepsilon + \sqrt{\beta_k} + c_1(\varepsilon)\beta_k)\|\xi_k\| + \int_0^t |\Delta x_k(s)|\,ds\right)$$

$$+\left|\int_0^t A_2^\varepsilon(s)A^\varepsilon(s)\,\Delta\psi_k(s)ds\right| \quad . \tag{18}$$

Taking into account (11),(12),(16),(17) and (18) we have

$$|\Delta x_k(t)| \leqslant |v_k - v_o| + c(\varepsilon + c_1(\varepsilon)\beta_k + |w_k'|\exp(-\sigma\frac{t}{\beta_k}) + |w_k' - w_o'|$$

$$+ (\varepsilon + \sqrt{\beta_k} + c_1(\varepsilon)\beta_k)(\|\Delta x_k\| + \|\Delta\psi_k\|) + \|\bar{y}_k - y_o\|$$

$$+ \int_0^t |\Delta x_k(s)|\,ds) + \left|\int_0^t \Delta\varphi_k(s)ds\right| + \left|\int_0^t A_2^\varepsilon(s)A^\varepsilon(s)\Delta\psi_k(s)ds\right|). \tag{19}$$

Let us remind that if the sequence $z_k(\cdot)$ of functions is L^2-weakly convergent to zero, then

$$\lim_{k\to\infty}\max_{0\leqslant t\leqslant 1}\left|\int_0^t z_k(s)ds\right| = 0 ,$$

and the sequence of L^2 norms $\|z_k\|$ is bounded. Applying the Gronwall lemma to (19) we get

$$\|\Delta x_k\| \leqslant \|\Delta x_k\|_C \leqslant c(\varepsilon + \sqrt{\beta_k} + c_1(\varepsilon)\beta_k)\|\Delta x_k\| + c .$$

Choosing $\varepsilon < 1/c$ and tending to zero with β_k we deduce

$$\limsup_{k\to\infty}\|\Delta x_k\| < +\infty \quad . \tag{20}$$

Using this in (19) gives us

$$|\Delta x_k(t)| \leqslant c(\varepsilon + \int_0^t |\Delta x_k(s)|\,ds + |w_k'| \exp(-\sigma \frac{t}{\beta_k}) + \delta_k). \qquad (21)$$

where $\lim_{k\to\infty} \delta_k = 0$ uniformly in $[0,1]$. From the Gronwall lemma we get that for every $\Theta \in (0,1)$

$$\lim_{k\to\infty} \max_{\Theta \leqslant t \leqslant 1} |\Delta x_k(t)| \leqslant c\varepsilon .$$

Since ε can be arbitrarily small and $\Delta x_k(\cdot)$ does not depend on ε the last relation implies (6).

The statement (ii) follows immediately from (21).

If $\lim_{k\to\infty} \beta_k\sqrt{w_k} = 0$, then from (6),(10),(12),(13) and (16)

$$\limsup_{k\to\infty} \|\Delta y_k\| < +\infty \quad .$$

Then, in order to prove (iii), it is sufficient to show that for every $t \in [0,1]$

$$\lim_{k\to\infty} \int_0^t \Delta y_k(s)\,ds = 0 . \qquad (22)$$

Using a sequence of inequalities similar to (18) we have

$$|\frac{1}{\beta_k} \int_0^t \int_0^s Y(s,v,\beta_k)\Delta\psi_k(v)\,dv\,ds| \leqslant c(\varepsilon + \sqrt{\beta_k} + c(\varepsilon)\beta_k)\|\Delta\psi_k\|$$
$$+ |\int_0^t A^{\varepsilon}(s)\Delta\psi_k(s)\,ds| \quad .$$

which, combined with (7) and (15) gives us (22).

The estimate for $\Delta x_k(\cdot)$ in (8) follows from (15),(19) and (20). In order to obtain the estimate for $\Delta y_k(\cdot)$ we use (12),(13) and (15).

Finally, let the conditions in (v) hold. Since $y_0(\cdot) \in C(R^n)$ one can find $y^{\delta}(\cdot) \in C^1(R^n)$ such that $\|y^{\delta} - y_0\|_C < \delta$. Then

$$|\bar{y}_k(t) - y_0(t)| \leqslant |\frac{1}{\beta_k} \int_0^t Y(t,s,\beta_k)A_4(s)(y_0(s) - y^{\delta}(s))\,ds|$$

$$+ \left| \int_0^t \frac{\partial}{\partial s} Y(t,s,\beta_k) y^{\delta}(s) ds - y_o(t) \right|$$

$$\leqslant c(\delta + |y^{\delta}(0)| \exp(-\sigma \frac{t}{\beta_k}) + \beta_k \| \dot{y}^{\delta} \|_C) \quad . \tag{23}$$

Substituting (23) in (12) we obtain (9). This proves the lemma completely.

Consider now the following singularly perturbed control system

$$\dot{x} = A_1(t)x + A_2(t)y + B_1(t)u \ , \ x(0) = x^o \ ,$$

$$\beta \dot{y} = A_3(t)x + A_4(t)y + B_2(t)u \ , \ y(0) = y^o \ , \tag{24}$$

on the interval $[0,1]$, assuming that the set of feasible controls is given by

$$U = \{ u(\cdot) \text{ - measurable } , \ u(t) \in V \text{ for a.e. } t \in [0,1] \}.$$

The corresponding reduced system is

$$\dot{x} = A_1(t)x + A_2(t)y + B_1(t)u \ , \ x(0) = x^o \ ,$$

$$0 = A_3(t)x + A_4(t)y + B_2(t)u \ . \tag{25}$$

We assume that A1 and the following condition hold:

A2. The matrix $B_1(t)$ is continuous, the matrices $A_2(t)$, $A_3(t)$, $A_4(t)$ and $B_2(t)$ are C^1, the set V is bounded.

In the sequel we denote by $\omega(u,\delta)_1$ the modulus of continuity of the function $u(\cdot)$ in $L^1(R^r)$, that is

$$\omega(u,\delta)_1 = \sup_{|h| \leqslant \delta} \int_{c(h)}^{d(h)} |u(t+h) - u(t)| \, dt \ ,$$

where $c(h) = 0, d(h) = 1-h$ for $h > 0$ and $c(h) = -h, d(h) = 1$ for $h < 0$.

<u>Lemma 3.2</u>. Let $u_\beta^1(\cdot)$ and $u_\beta^2(\cdot)$ be feasible controls, $(x_\beta^1(\cdot), y_\beta^1(\cdot))$ solve (24) for $u_\beta^1(\cdot)$ and $(x_\beta^2(\cdot), y_\beta^2(\cdot))$ satisfy (25) for $u_\beta^2(\cdot)$. Then there exists a constant c which does not depend on $u_\beta^1(\cdot)$ and $u_\beta^2(\cdot)$

such that

$$\|x_\beta^1 - x_\beta^2\|_C \leqslant c(\beta + T(\beta)),$$

$$\|y_\beta^1 - y_\beta^2\|_{L^1} \leqslant c(\beta + T(\beta) + \omega(u_\beta^2,\beta)_1),$$

where

$$T(\beta) = \text{meas}\left\{ t\in[0,1],\ u_\beta^1(t) \neq u_\beta^2(t)\right\}.$$

Proof. Denote

$$\xi_\beta(t) = A_3(t)(x_\beta^1(t) - x_\beta^2(t)) + B_2(t)(u_\beta^1(t) - u_\beta^2(t)),$$

$$\eta_\beta(t) = \frac{1}{\beta}\int_0^t Y(t,s,\beta)\xi_\beta(s)ds.$$

Clearly, $\eta_\beta(\cdot)$ is uniformly bounded when $\beta\to 0$.By repeating the arguments in (18) one can get

$$\left|\int_0^t A_2(s)\eta_\beta(s)ds\right| \leqslant c_1(\beta + \int_0^t |\xi_\beta(s)|ds) \tag{26}$$

and

$$\|\eta_\beta\|_{L^1} \leqslant c_2(\beta + T(\beta) + \|x_\beta^1 - x_\beta^2\|_C). \tag{27}$$

Moreover, using (10) and the boundedness of $y_\beta^2(\cdot)$ we have

$$- \int_0^1 A_2(t)(\frac{1}{\beta}\int_0^t Y(t,s,\beta)A_4(s)y_\beta^2(s)ds + y_\beta^2(t))dt$$

$$= \int_0^1 A_2(t)A_4^{-1}(t)\frac{d}{dt}\int_0^t Y(t,s,\beta)A_4(s)y_\beta^2(s)dsdt$$

$$\leqslant |A_2(1)A_4^{-1}(1)\int_0^1 Y(1,s,\beta)A_4(s)y_\beta^2(s)ds|$$

$$+ \int_0^1 \left|\frac{d}{dt}(A_2(t)A_4^{-1}(t))\int_0^t Y(t,s,\beta)A_4(s)y_\beta^2(s)ds\right|dt = O(\beta).$$

Using this estimate and (26) in a relation for the difference $x_\beta^1(\cdot) - x_\beta^2(\cdot)$ analogical to (11) and applying the Gronwall lemma we obtain the estimate for the slow states.Furthermore

$$|\frac{1}{\beta}\int_0^t Y(t,s,\beta)A_4(t)y_\beta^2(t)ds + y_\beta^2(t)| \leq c\beta \tag{28}$$

for all $t\in[0,1]$. Denoting $g_\beta(t) = A_4(t)y_\beta^2(t)$, $g_\beta(t) = 0$ for $t>0$ we have

$$\int_0^1 |\frac{1}{\beta}\int_0^t Y(t,s,\beta)A_4(s)y_\beta^2(s)ds + y_\beta^2(t)|dt$$

$$\leq \int_0^1 |\frac{1}{\beta}\int_0^t Y(t,s,\beta)(g_\beta(s) - g_\beta(t))ds|dt$$

$$+ \int_0^1 |\frac{1}{\beta}\int_0^t Y(t,s,\beta)A_4(t)y_\beta^2(t)ds + y_\beta^2(t)|dt , \tag{29}$$

and

$$\frac{1}{\beta}\int_0^1 |\int_0^t Y(t,s,\beta)(g_\beta(s) - g_\beta(t))ds|dt$$

$$\leq \frac{\sigma_0}{\beta}\int_0^1 \int_0^t \exp(-\sigma\frac{t-s}{\beta})|g_\beta(s) - g_\beta(t)|dsdt$$

$$\leq \sigma_0 \int_0^1 \int_0^{t/\beta} e^{-\sigma s}|g_\beta(t-\beta s) - g_\beta(t)|dsdt \leq \sigma_0 \int_0^{1/\beta} e^{-\sigma s}\omega(g_\beta,\beta s)_1 ds$$

$$+ O(\beta) \leq \sigma_0 \sum_{k=1}^{p} e^{-\sigma k}\omega(g_\beta,(k+1)\beta)_1 + O(\beta) ,$$

where $p > 1/\beta + 1$, p - integer ,

$$\leq \sigma_0\omega(g_\beta,\beta)_1 \sum_{k=1}^{\infty} e^{-\sigma k}(k+1) + O(\beta) \leq c_3(\omega(g_\beta,\beta)_1 + \beta). \tag{30}$$

Clearly

$$\omega(g_\beta,\beta)_1 \leq c_4(\omega(u_\beta^2,\beta)_1 + \beta) . \tag{31}$$

Combining (28)-(31) we obtain finally

$$\int_0^1 \left| \frac{1}{\beta} \int_0^t Y(t,s,\beta) A_4(s) y_\beta^2(s) ds + y_\beta^2(t) \right| dt \leqslant c_5(\omega(u_\beta^2,\beta)_1 + \beta). \quad (32)$$

Then the second estimate follows from (12),(27) and (32),Q.E.D.

Consider now the reduced system

$$\dot{x} = A_0(t)x + B_0(t)u \ , \ x(0) = x^0 \ , \quad (33)$$

assuming that:

A3. The components of $A_0(t)$ and $B_0(t)$ are in $L^1(R^1)$.The admissible set of controls U consists of all integrable functions $u(\cdot)$ with values in a convex set $V \subset R^r$ for a.e. $t \in [0,1]$.

Let us remind that the reachable set P at the time t = 1 for the system (33) consists of all point in R^m which can be achieved at the time t = 1 begining from the initial state x^0 at the time t = 0, using controls from the admissible set U.Obviously, the set P is convex.

Lemma 3.3. Let x^1 belong to the relative interior of P.There exist numbers $\varepsilon_0 > 0$ and $c > 0$ such that for every $\varepsilon \in (0,\varepsilon_0)$ and for every x', $x'' \in P$, $|x' - x^1| < \varepsilon, |x'' - x^1| < \varepsilon$,if the feasible control $u'_\varepsilon(\cdot)$ drives the state from x^0 to x' at the time t = 1, there exists a feasible control $u''_\varepsilon(\cdot)$, which drives the state from x^0 to x'' at the time t = 1, so that the set

$$T(\varepsilon) = \{ t \in [0,1] \ , \ u'_\varepsilon(t) \neq u''_\varepsilon(t) \}$$

consists of no more than m+1 intervals $\Delta_1^\varepsilon, \ldots, \Delta_{m+1}^\varepsilon$, meas $\bigcup_{i=1}^{m+1} \Delta_i^\varepsilon < c\varepsilon$, and

$$\omega(u''_\varepsilon,\delta)_1 \leqslant \omega(u'_\varepsilon,\delta)_1 + c\varepsilon$$

for $\delta \in (0,\varepsilon]$.

Proof. For simplicity, let int $P \neq \emptyset$ .Otherwise, one can restrict all the further considerations to the affine hull of P.

There exists a simplex $S \subset P$, such that $x^1 \in$ int S.Each of the vertices x_s of S can be reached by means of a feasible control $u_s(\cdot)$. Let $u_s^h(\cdot)$ be a piecewise constant approximation of $u_s(\cdot)$ with a step lenght h.Obviously, the corresponding final state x_s^h converges to

x_s when $h \to 0$.Then for some sufficiently small but fixed h the point x^1 belongs to the interior of the simplex S^h with vertices x_s^h.Then there exists $\varepsilon_1 > 0$ such that if $|x - x^1| < \varepsilon_1$,then x is a convex combinations of x_s^h.Hence,every x in the ε_1-neighbourhood of x^1 can be reached by means of a convex combination of m+1 piecewise constant functions.We proved that there exists a ε_1-neighbourhood of x^1,every point of which can be reached by a control of variation bounded uniformly in this ε_1-neighbourhood.

Let S^m be the unit sphere in R^m and let $\varepsilon_0 \in (0, \varepsilon_1)$ be fixed.There exists $\alpha > 0$ such that for every $\varepsilon \in (0, \varepsilon_0)$, $x \in R^m$, $|x - x^1| < \varepsilon$ and $l \in S^m$ the point $x + \alpha l$ satisfies $|x + \alpha l - x^1| < \varepsilon_1$.Let $\varepsilon \in (0, \varepsilon_0)$, $x'_\varepsilon \in R^m$, $x''_\varepsilon \in R^m$, $|x'_\varepsilon - x^1| < \varepsilon$, $|x''_\varepsilon - x^1| < \varepsilon$, be arbitrarily chosen and let the feasible control $u'_\varepsilon(\cdot)$ correspond to x'_ε.Gor a given $l \in S^m$ one can find a control $u_l(\cdot) \in U$ of bounded variation uniformly in the ε_0-neighbourhood of x^1,which drives the state of (33) to $x' + \alpha l$. Denoting $\Delta u_l(\cdot) = u_l(\cdot) - u'_\varepsilon(\cdot)$ we have

$$\int_0^1 E(1,t)B_0(t)\Delta u_l(t)dt = \alpha l \ ,$$

where E(t,s) is the fundamental matrix solution of (33) normalized at t=s.Hence

$$\int_0^1 l^T E(1,t)B_0(t)\Delta u_l(t)dt = \alpha \quad . \tag{34}$$

Let $\delta = \min\{6\varepsilon/\alpha, 1\}$.We show that there exists an interval $\Delta_l^\varepsilon \subset [0,1]$ such that meas $\Delta_l^\varepsilon \leqslant \delta$ and

$$\int_{\Delta_l^\varepsilon} l^T E(1,t)B_0(t)\Delta u_l(t)dt \geqslant 3\varepsilon \ . \tag{35}$$

Let $\Delta_1, \ldots, \Delta_p$ be a covering of $[0,1]$ such that meas $\Delta_i \leqslant \delta$,where the integer $p > 1/\delta + 1$.If (35) does not hold for every Δ_i then

$$\int_0^1 l^T E(1,t)B_0(t)\Delta u_l(t)dt < 3p\varepsilon \leqslant 3(1/\delta + 1)\varepsilon \leqslant 6\varepsilon/\delta \leqslant \alpha \ ,$$

which contradicts (34).Thus, (35) holds.

Introduce the control

$$u_l^{\varepsilon}(t) = \begin{cases} u_{\varepsilon}'(t) & \text{for } t \in [0,1] \setminus \Delta_l^{\varepsilon} , \\ u_l(t) & \text{for } t \in \Delta_l^{\varepsilon} \quad . \end{cases}$$

Obviously $u_l^{\varepsilon}(\cdot) \in U$ and the corresponding final state x_l^{ε} from (33) satisfies

$$l^T(x_l^{\varepsilon} - x_{\varepsilon}') \geqslant 3\varepsilon \quad . \tag{36}$$

Define the set

$$W_{\varepsilon} = \text{co}\left\{ x_l^{\varepsilon},\ l \in S^m \right\} .$$

From (36) it follows that for every $l \in S^m$

$$\sup_{x \in W_{\varepsilon}} l^T(x - x_{\varepsilon}') \geqslant 3\varepsilon . \tag{37}$$

We prove that

$$B_{\varepsilon} = \left\{ x \in R^m,\ |x - x_{\varepsilon}'| < 2\varepsilon \right\} \subset W_{\varepsilon} . \tag{38}$$

Assume that there exists $\bar{x} \in B_{\varepsilon}$ such that $\bar{x} \notin W_{\varepsilon}$. Since W_{ε} is convex, by the separation theorem there exists $\bar{l} \in S^m$ such that

$$\bar{l}^T x \leqslant \bar{l}^T \bar{x} \quad \text{for every } x \in W_{\varepsilon} .$$

Then

$$\sup_{x \in W_{\varepsilon}} \bar{l}^T(x - x_{\varepsilon}') \leqslant \bar{l}^T(\bar{x} - x_{\varepsilon}')$$

and taking advantage of (37) we obtain

$$3\varepsilon \leqslant \bar{l}^T(\bar{x} - x_{\varepsilon}') \leqslant |\bar{x} - x_{\varepsilon}'| < 2\varepsilon \quad .$$

This contradiction gives us (38). This, since $x_{\varepsilon}'' \in B_{\varepsilon}$, then $x_{\varepsilon}'' \in W_{\varepsilon}$. This means that x_{ε}'' can be reached by means of a control $u_{\varepsilon}''(\cdot)$, which is a convex combination of no more than m+1 controls from the set $\left\{ u_l(\cdot),\ l \in S^m \right\}$. The function $\Delta u(\cdot) = u_{\varepsilon}'(\cdot) - u_{\varepsilon}''(\cdot)$ differs from zero on a union of no more than m+1 intervals Δ_i^{ε} with measure less than $(m+1)\delta \leqslant 6\varepsilon(m+1)/\alpha$ .On each interval the variation of $\Delta u(\cdot)$ is bounded by a constant which depends on ε_o only.By the known estimate

$$\omega(g,\gamma)_{L^1(0,\beta)} \leqslant \gamma V_0^{\beta}(g) \text{ for } \gamma \in [0,\beta]$$

we conclude that there exists a constant c_1 such that $\omega(\Delta u,\delta)_1 \leqslant c_1\varepsilon$. Taking $c = \max\{c_1, 6(m+1)/\alpha\}$ we complete the proof.

3.3.Convergence of the reachable sets

We begin with an example:

Example 3.1. Consider the system

$$\beta\dot{y}_1 = -y_1 + u \ , y_1(0) = 0,$$
$$\beta\dot{y}_2 = -2y_2 + u, y_2(0) = 0,$$
$$u(t) \in [-1,1] \text{ for a.e. } t \in [0,1] \ ,$$
$$u(\cdot) \text{ - measurable, } \beta > 0.$$

For $\beta = 0$ we get algebraic equations, that is

$$0 = -y_1 + u \ ,$$
$$0 = -2y_2 + u$$

and the state becomes a function with the same properties as the control.Thus, one can not define formally a reachable set for the reduced system.

We can slightly change the admissible set of controls taking

$$u(t) = \begin{cases} v(t) & \text{for } t \in [0,1) \ , \\ w & \text{for } t = 1 \ , \end{cases}$$

where $v(\cdot)$ is measurable, $v(t) \in [-1.1]$ for a.e. $t \in [0,1)$,and $w \in [-1,1]$. Clearly,such a change does not affect the reachable set at t=1 of the perturbed system with $\beta > 0$.Then the final state $(y_1(1),y_2(1))$ of the reduced system is well-defined and the segment

$$KR = \{ (y_1,y_2) \ , \ y_1 \in [-1,1] \ , \ y_2 = 0.5y_1 \}$$

is exactly the reachable set of the reduced system. In other words, the set KR is obtained by taking $\beta=0$ and $t=1$ in our system.

Apply the feasible control

$$u^{\beta}(t) = \begin{cases} 1 & \text{for } t\in[0,1+\beta\ln 0.5) \ , \\ -1 & \text{for } t\in[1+\beta\ln 0.5,1] \end{cases}$$

to the perturbed system. We get

$$y_1^{\beta}(1) = e^{-1/\beta} \ , \quad y_2^{\beta}(1) = -0.25 - 0.5e^{-2/\beta} \ .$$

Then

$$\lim_{\beta\to 0} (y_1^{\beta}(1), y_2^{\beta}(1)) = (0, -0.25) \notin KR \ ,$$

that is KR does not contain all limit points of sequences from the perturbed reachable set when $\beta\to 0$. This means that the multivalued mapping $\beta \to$ "reachable set" is discontinuous at $\beta=0$.

Observe that the boundedness of the control is not essential for such a conclusion. If we define the feasible controls as

$$u(t) = \begin{cases} v(t) \in R^1 & \text{for } t\in[0,1) \ , \\ w \in R^1 & \text{for } t=1 \ , \end{cases}$$

where $v(\cdot)$ is integrable, The set KR will be the line

$$\{(y_1,y_2) \ , \ y_1 = 2y_2 \ , \ y_2\in R^1\} \ ,$$

which does not contain the point (0, 0.25).

In this section we show that the reachable sets of singularly perturbed control systems possess Hausdorff limits at $\beta=0$ which contain but does not coincide with the sets obtained by formal substituting $\beta=0$, as the set KR above. This enables us to establish well-posedness of singularly perturbed optimal control problems.

We consider singularly perturbed linear systems

$$\begin{aligned} \dot{x} &= A_1(t)x + A_2(t)y + B_1(t)u \ , \ x(0) = x^0 \ , \\ \beta\dot{y} &= A_3(t)x + A_4(t)y + B_2(t)u \ , \ y(0) = y^0 \ , \end{aligned} \qquad (39)$$

with an admissible set of controls

$$U = \{u(\cdot) \in L^1(R^r),\ u(t) \in V \text{ for a.e. } t \in [0,1]\},$$

assuming that:

A3. The matrices $A_i(t)$ and $B_j(t)$ are continuous. The eigenvalues of the matrix $A_4(t)$ have negative real parts for all $t \in [0,1]$. The set V is compact.

For every $x \in R^m$ we define the set

$$R(x) = - A_4^{-1}(1)A_3(1) + R ,$$

where

$$R = \int_0^{\infty} \exp(A_4(1)s)B_2(1)V ds .$$

Here the integral of the set-valued mapping is taken in the sense of Aumann [1], that is

$$R = \{ \int_0^{\infty} \exp(A_4(1)s)B_2(1)v(s)ds ,\ v(s) \in V \text{ for a.e. } t \in [0,+\infty) ,$$

$$v(\cdot) - \text{integrable} \} .$$

The assumption A3 yields that R is a properly defined compact and convex set in R^n, see Joffe and Tikhomirov [42], p.349. Observe that R is exactly the closure of the 0-controllable set of the time-reversed system

$$\dot{y} = - A_4(1)y - B_2(1)u ,$$

that is, the set of all points in R^n which can be driven at a finite time to the origin by means of integrable controls with values in the set V.

Denote by K_β the reachable set at t=1 for the perturbed system (39), $K_\beta \subset R^{m+n}$. Let P be the reachable set at t=1 for the reduced system

$$\dot{x} = A_o(t)x + B_o(t)u ,\ x(0) = x^o \qquad (33)$$

$$A_o = A_1 - A_2A_4^{-1}A_3 ,\ B_o = B_1 - A_2A_4^{-1}B_2 ,$$

obtained for $\beta = 0$. Observe that taking co V instead of V does not change the sets R, K_β and P. Thus, the admissible set of controls U can be thought of as a L^2 - weakly compact set.

Let

$$K_o = \{(x,y) \in R^{m+n} \ , \ x \in P \ , \ y \in R(x)\} .$$

We prove that:

<u>Theorem 3.1</u>. Let $d_H(K_\beta, K_o)$ be the Hasdorff distance between the sets K_β and K_o. Then

$$\lim_{\beta \to 0} d_H(K_\beta, K_o) = 0 .$$

Proof. Clearly, K_β and K_o are convex and compact sets. Let (x_o, y_o) be from K_o. There exists a control $u_o(\cdot) \in U$ such that the corresponding solution of (33) satisfies $x_o(1) = x_o$, and there exists an integrable function $v_o(\cdot)$, $v_o(t) \in V$ for $t \in [0, +\infty)$ such that

$$y_o = - A_4^{-1}(1)A_3(1)x_o + \int_0^\infty \exp(A_4(1)s)B_2(1)v_o(s)ds .$$

Define the control

$$u_\beta(t) = \begin{cases} u_o(t) & \text{for } t \in [0, 1 - \beta] , \\ v_o\left(\frac{1-t}{\beta}\right) & \text{for } t \in (1 - \beta, 1] . \end{cases}$$

Then $\lim_{\beta \to 0} u_\beta(t) = u_o(t)$ for almost all $t \in [0,1]$. Let $(x_\beta(\cdot), y_\beta(\cdot))$ be the solution of the perturbed system (39) which results from $u_\beta(\cdot)$. By Lemma 3.1(i) it follows that $\lim_{\beta \to 0} x_\beta(1) = x_o$. Denote by $\bar{y}_\beta(\cdot)$ the solution of the equation

$$\beta\dot{y} = A_4(1)y + A_3(1)x_\beta(1) + B_2(1)u_\beta(t), \qquad (40)$$

$$y(0) = y^o ,$$

and let $\Delta y_\beta = y_\beta - \bar{y}_\beta$, $\Delta x_\beta(t) = x_\beta(t) - x_\beta(1)$, $\Delta A_i(t) = A_i(t) - A_i(1)$, $i=3,4$, $\Delta B_2(t) = B_2(t) - B_2(1)$. We have

$$\beta\Delta\dot{y}_\beta = A_4(1)\Delta y_\beta + \Delta A_4(t)y_\beta(t) + \Delta A_3(t)x_\beta(t)$$

$$+ A_3(1)\Delta x_\beta(t) + \Delta B_2(t)u_\beta(t) \ , \quad \Delta y_\beta(0) = 0. \tag{41}$$

Using the compactness of V, (10) and the Gronwall lemma one can easily prove that $y_\beta(t)$, $\bar{y}_\beta(t)$ and $x_\beta(t)$ are bounded uniformly in t and β, hence $x_\beta(t)$ is Lipschitz continuous with respect to t uniformly in β.From (10) and (41) we get

$$|\Delta y_\beta(1)| \leqslant \frac{\sigma_0}{\beta}\int_0^{1-\sqrt{\beta}} \exp(-\sigma\frac{1-t}{\beta})|\Delta A_4(t)y_\beta(t) + \Delta A_3(t)x_\beta(t)$$

$$+ A_3(1)\Delta x_\beta(t) + \Delta B_2(t)u_\beta(t)|dt$$

$$+ \max_{1-\sqrt{\beta}\leqslant t\leqslant 1} (|\Delta A_4(t)y_\beta(t)| + |\Delta A_3(t)x_\beta(1)|$$

$$+ |A_3(1)\Delta x_\beta(t)| + |\Delta B_2(t)u_\beta(t)|)\frac{\sigma_0}{\beta}\int_{1-\sqrt{\beta}}^{1} \exp(-\sigma\frac{1-t}{\beta})dt.$$

Hence

$$\lim_{\beta\to 0} |\Delta y_\beta(1)| = 0. \tag{42}$$

We have

$$\lim_{\beta\to 0} \frac{1}{\beta}\int_0^1 \exp(A_4(1)\ \frac{1-t}{\beta})A_3(1)x_\beta(1)dt = - A_4^{-1}(1)A_3(1)x_o \tag{43}$$

and

$$\frac{1}{\beta}\int_0^1 \exp(A_4(1)\ \frac{1-t}{\beta})B_2(1)u_\beta(t)dt$$

$$= \frac{1}{\beta}\int_0^{1-\sqrt{\beta}} \exp(A_4(1)\ \frac{1-t}{\beta})B_2(1)u_o(t)dt$$

$$+ \int_0^{1/\sqrt{\beta}} \exp(A_4(1)s)B_2(1)v_o(s)ds \ . \tag{44}$$

Applying (42), (43) and (44) to the Cauchy formula for (40) we conclude that

$$\lim_{\beta\to 0} |y_\beta(1) - y_o| = 0.$$

Hence, there exists a sequence $(x_\beta, y_\beta) \in K_\beta$ such that

$$\lim_{\beta\to 0} (x_\beta, y_\beta) = (x_o, y_o) . \tag{45}$$

Now, let us assume that there exist sequences β_k, $\lim_{k\to\infty} \beta_k = 0$, $(x_k, y_k) \in K_{\beta_k}$ such that $\lim_{k\to\infty} (x_k, y_k) \notin K_o$. Let the control $u_k(\cdot)$ correspond to (x_k, y_k). The sequence $u_k(\cdot)$ has a weak limit point $u_o(\cdot)$ in $L^2(R^r)$ and let $x_o(\cdot)$ result from (33) for $u_o(\cdot)$. Then, by Lemma 3.1 (ii), $x_o = x_o(1) \in P$. As before we denote by Δy_k the difference $y_k - \bar{y}_k$, where $\bar{y}_k(\cdot)$ is the solution of (40) for $u_k(\cdot)$ and β_k. By repeating the arguments in (42) we have $\lim_{k\to\infty} \Delta y_k(1) = 0$ and

$$y_k(1) = - A_4^{-1}(1)A_3(1)x_o + \int_0^{1/\beta_k} \exp(A_4(1)s)B_2(1)u_k(1-\beta_k s)ds$$

$$+ \varphi_k^1 \in R(x_o) + \varphi_k^2 ,$$

where

$$\lim_{k\to\infty} (|\varphi_k^1| + |\varphi_k^2|) = 0.$$

Hence $y_o \in R(x_o)$ and $(x_o, y_o) \in K_o$, which is a contradiction, This, combined with (45) gives us Hausdorff convergence, Q.E.D.

<u>Example 3.1</u>(Continuation). The boundary of the reachable set K_β for the system

$$\beta\dot{y}_1 = - y_1 + u , \quad y_1(0) = 0 ,$$

$$\beta\dot{y}_2 = -2y_2 + u , \quad y_2(0) = 0 ,$$

can be achieved by means of bang-bang controls having one switching. Using the switching point t_β as a patameter one can deduce that every point $(y_1, y_2) \in K_\beta$ satisfies

$$y_1 = \pm 2\exp\left(\frac{t_\beta - 1}{\beta}\right) + O(\beta) ,$$

$$y_2 = \pm(\exp(2(t_\beta - 1)/\beta) - 0.5) + O(\beta) .$$

Eliminating t_β and letting β tend to zero we obtain that

$$K_o = R = \{y = (y_1, y_2) \in R^2 , \quad -1 \leqslant y_1 \leqslant 1,$$

$$0.25(y_1 + 1)^2 - 0.5 \leqslant y_2 \leqslant - 0.25(y_1 - 1)^2 + 0.5 \} .$$

The sets R and KR are given in Fig.3.1

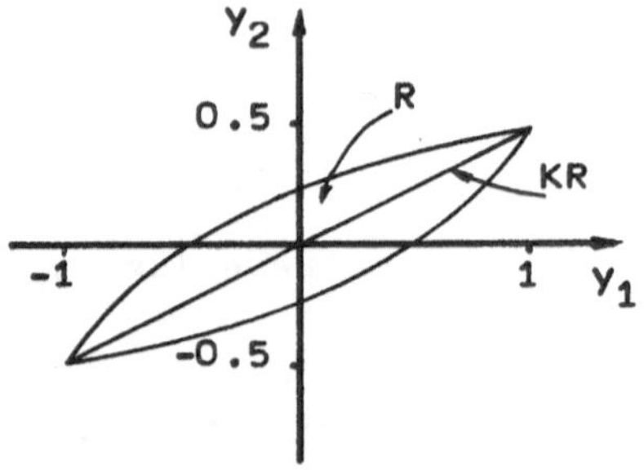

Fig.3.1

Remark 3.1. The result, obtained in Theorem 3.1 may be interpreted in the following way: Let $Z_\beta(t)$ be the family of the solutions of the differential inclusion

$$\beta\dot{y} \in A_4(t)y + B_2(t)V \quad , \; y(0) = y_o \quad , \; t \in [0,1] .$$

Define the set

$$Z_o(t) = \int_0^\infty \exp(A_4(t)s)B_2(t)Vds .$$

Then, for every $t \in (0,1]$

$$\lim_{\beta \to 0} d_H(Z_\beta(t), Z_o(t)) = 0.$$

Furthermore

$$-A_4^{-1}(t)B_2(t)V \subsetneq Z_o(t) ,$$

that is, the set $Z_\beta(t)$ does not converge to the set of solutions of the generalized equation

$$0 \in A_4(t)y + B_2(t)V \ ,$$

obtained for $\beta = 0$.In other words, the classical Tichonov's theorem for singularly perturbed differential equations, see Tichonov [70] , can not be formally extended to differential inclusions.

Let us consider the case of unrestricted controls, that is the perturbed system system (39) with matrices as in A3 and $V = R^r$. The admissible set of controls is $L^2(R^r)$.Let K_β be the reachable set at t=1 of (39) and let P denote the reachable set at t=1 of the reduced system (33).Clearly, K_β and P are subspaces.

Denote by $A_4(1)/B_2(1)$ the controllable subspace for the system

$$\dot{y} = A_4(1)y + B_2(1)u \ .$$

Let

$$K_o = \{(x,y) \in R^{m+n} \ , \ x \in P, \ y \in -A_4^{-1}(1)A_3(1)x + A_4(1)/B_2(1)\} \ .$$

The following result is to a certain extend an analogue of Theorem 3.1.

Theorem 3.2. Let $z \in K_o$ and the sequence β_k, $\lim \beta_k = 0$ as $k \to \infty$ be arbitrarily chosen.Let the control $u(\cdot)$ drive the state of the reduced system (33) along $x(\cdot)$ from x^o at t=0 to the projection of z on R^m at t=1.Then there exists a sequence of controls $u_k(\cdot)$, pointwise convergent to $u(\cdot)$ on (0,1), such that the corresponding trajectory of the perturbed system (39) with β_k satisfies

$$\lim_{k \to \infty} (x_k(1), y_k(1)) = z$$

and

$$\lim_{k \to \infty} (\|x_k - x\|_C + \|y_k + A_4^{-1}(A_3x + B_2u)\|) = 0 \ .$$

Proof. Let $\varepsilon_k \to 0$ and $\lim_{k \to \infty} \varepsilon_k / \beta_k = +\infty$.Define the control

$$\tilde{u}_k(t) = \begin{cases} u(t) & \text{for } t \in [0, 1-\varepsilon_k] \\ 0 & \text{for } t \in (1-\varepsilon_k, 1] \end{cases} \ .$$

Let $(\tilde{x}_k(\cdot),\tilde{y}_k(\cdot))$ correspond to $\tilde{u}_k(\cdot)$ and β_k according to (39). Then from Lemma 3.1(ii) and (iv) we get

$$\lim_{k\to\infty} (\|\tilde{x}_k - x\|_C + \|\tilde{y}_k - y\|) = 0 ,$$

where $y(t) = - A_4^{-1}(t)(A_3(t)x(t) + B_2(t)u(t))$.Furthermore

$$\tilde{y}_k(1) = Y(1,0,\beta_k)y^o + \frac{1}{\beta_k}\int_0^1 Y(1,t,\beta_k)A_3(t)\tilde{x}_k(t)dt$$
$$+ \frac{1}{\beta_k}\int_0^{1-\varepsilon_k} Y(1,t,\beta_k)B_2(t)u(t)dt .$$

Clearly

$$\lim_{k\to\infty}\frac{1}{\beta_k}\int_0^{1-\varepsilon_k} Y(1,t,\beta_k)B_2(t)u(t)dt = 0.$$

Choose an arbitrary $\delta>0$ and a matrix A^δ with smooth components such that $\|A^\delta - A_4^{-1}A_3\|_C < \delta$.Proceeding in the same way as in the proof of Lemma 3.1 one can get the estimate

$$\left|\frac{1}{\beta_k}\int_0^1 Y(1,t,\beta_k)A_3(t)\tilde{x}_k(t)dt + A_4^{-1}(1)A_3(1)v\right|$$

$$\leqslant c(\delta + c_1(\delta)\beta_k + O(\beta_k)) ,$$

where v is the projection of z on R^m,i.e. $z = (v,w)$ and $c_1(\delta) \geqslant \|\frac{d}{dt}A^\delta\|$ .The last two relations yield

$$\lim_{k\to\infty} \tilde{y}_k(1) = - A_4^{-1}(1)A_3(1)v \quad . \tag{46}$$

Introduce the matrices

$$M_k = \frac{1}{\beta_k}\int_{1-\sqrt{\beta_k}}^1 Y(1,t,\beta_k)B_2(t)B_2^T(t)Y(1,t,\beta_k)^T dt ,$$

$$\tilde{M}_k = \frac{1}{\beta_k}\int_{1-\sqrt{\beta_k}}^1 \exp(A_4(1)\,\frac{1-t}{\beta_k})B_2(1)B_2^T(1)\exp(A_4^T(1)\,\frac{1-t}{\beta_k})dt ,$$

$$M = \int_0^\infty \exp(A_4(1)t)B_2(1)B_2^T(1)\exp(A_4^T(1)t)dt \ .$$

Using (10) and the continuity of $A_4(\cdot)$ and $B_2(\cdot)$ at $t=1$ we get

$$\lim_{k\to\infty} |M_k - \tilde{M}_k| = 0.$$

Furthermore

$$|\tilde{M}_k - M| \leq c \int_{1/\sqrt{\beta_k}}^\infty \exp(-2\sigma s)ds = O(\beta_k) \ ,$$

which yields

$$\lim_{k\to\infty} M_k = M \ . \tag{47}$$

In Wonham [80],p. 64 , it is proved that

$$A_4(1)/B_2(1) = \text{im } M \ .$$

Since $w + A_4^{-1}(1)A_3(1)v \in A_4(1)/B_2(1)$ one can find $g \in R^n$ such that

$$Mg = w + A_4^{-1}(1)A_3(1)v \ .$$

Define the control

$$u_k(t) = \begin{cases} \tilde{u}_k(t) & \text{for } t \in [0, 1 - \sqrt{\beta_k}) \ , \\ B_2^T(t)Y(1,t,\beta_k)g & \text{for } t \in [1 - \sqrt{\beta_k} \ ,1]. \end{cases}$$

Applying $u_k(\cdot)$ to the perturbed system (39) with β_k we get the trajectories $x_k(\cdot)$ and $y_k(\cdot)$, for which

$$\lim_{k\to\infty} \|x_k - x\|_C = 0$$

and

$$y_k(1) - \tilde{y}_k(1) = \frac{1}{\beta_k} \int_0^1 Y(1,t,\beta_k)A_3(t)(x_k(t) - \tilde{x}_k(t))dt$$

$$+ Mg + (M_k - M)g \ .$$

Thus, in view of (46) and (47)

$$\lim_{k\to\infty} y_k(1) = \lim_{k\to\infty} \tilde{y}_k(1) + Mg = w .$$

Observe that the sequence $u_k(\cdot)$ converges pointwise to $u(\cdot)$ and

$$\lim_{k\to\infty} \|u_k - u\| = 0 .$$

Then the convergence of $y_k(\cdot)$ follows from Lemma 3.1 (iv), Q.E.D.

If

$$\operatorname{rank}\left[B_2(1), A_4(1)B_2(1), \ldots, A_4^{n-1}(1)B_2(1)\right] = n ,$$

then $A_4(1)/B_2(1) = R^n$, hence $K_o = P \times R^n$. This particular result was obtained in Dontchev and Gičev [18].

3.4. Well-posedness

3.4.1. Mayer's problem

This paragraph is concerned with the well-posedness of the order reduction procedure for the following optimal control problem

$$g(x,y) \longrightarrow \inf \quad , \ (x,y) \in K_\beta \ , \tag{48}$$

where, as before, K_β is the reachable set at $t=1$ of the system

$$\dot{x} = A_1(t)x + A_2(t)y + B_1(t)u \ , x(0) = x^o \ ,$$
$$\beta\dot{y} = A_3(t)x + A_4(t)y + B_2(t)u \ , y(0) = y^o \ , \tag{49}$$

with an admissible set of controls

$$U = \left\{u(\cdot) \in L^1(R^r),\ u(t) \in V \text{ for a.e. } t \in [0,1]\right\}. \tag{50}$$

We assume that the condition A3 from the previous section holds and

A4. The function $g(\cdot)$ is continuous.

Since the set K_β is compact, there exists an optimal control $\hat{u}_\beta(\cdot)$ which, when applied to (49), drives the initial state (x^o, y^o)

to $(\hat{x}_\beta,\hat{y}_\beta)$ minimizing the function $g(\cdot)$.

In view of the results of the previous section we define the limit problem as

$$g(x,y) \longrightarrow \inf \ , \ (x,y) \in K_o \ , \tag{51}$$

where the compact set K_o is determined as

$$K_o = \left\{ (x,y) \in R^{m+n}, \ x \in P \ , \ y \in R(x) \right\} \ ,$$

$$R(x) = - A_4^{-1}(1)A_3(1)x + R \ ,$$

$$R = \int_0^\infty \exp(A_4(1)s)B_2(1)Vds \ ,$$

and P is the reachable set at t=1 for the reduced system

$$\dot{x} = A_o(t)x + B_o(t)u \ , \ x(0) = x^o \ . \tag{52}$$

Notice that the limit problem does not depend on the initial condition y^o.

The optimal control (which exists) for the reduced problem (51) is denoted by $\hat{u}_o(\cdot)$ and the optimal final state by $(\hat{x}_o,\hat{y}_o)$.

We can rewrite the limit problem as follows

$$g_o(x) \longrightarrow \inf \ , \ x \in P \ , \tag{53}$$

where the (continuous) function $g_o(\cdot)$ is defined by

$$g_o(x) = \inf_{y \in R} g(x,y - A_4^{-1}(1)A_3(1)x) \ . \tag{54}$$

Such a definition suggests that the limit problem can be solved as a two-stage optimization problem.The goal function $g_o(x)$ of the "outher" problem (53) is to be evaluated by means of the "inner" problem (54).Clearly,in so far as the set R can be effectively approximated , the inner problem is a parametric mathematical programming problem.Generally, the problem (54) can be regarded as a parametric Mayer's problem with objective function $g(x,y - A_4^{-1}(1)A_3(1)x)$ for the system

$$\dot{y} = A_4(1)y + B_2(1)u \ , \ y(0) = 0 \ ,$$

on infinite time horizon, the reachable set of which is exactly R. The outher problem (53) for the slow state x remains Mayer's problem over [0,1].We shall not go into computational details further noting only that if g(x,y) is separable, one can select two important cases: 1) the function $g(x,\cdot)$ is linear, and 2) $A_3(1) = 0$.If one of these conditions holds, the problems (53) and (54) can be solved independently.The solution of (54) gives only a shift constant for $g_0(x)$, which actually provides well-posedness.

It is clear that if the function $g(\cdot)$ does not depend explicitly on the fast states y, the discontinuity of the reachable set at $\beta=0$,described in the previous section,does not affect the well-posedness.Such problems were considered by Dmitriev [11] for a system linear with respect to the state, and by Binding [4] for a nonlinear system.

From Theorem 3.1 and the continuity of the function $g(\cdot)$ we immediately obtain

Theorem 3.3. The problem (48) is well-posed in the sense of performance convergence, that is

$$\lim_{\beta \to 0} g(\hat{x}_\beta, \hat{y}_\beta) = g(\hat{x}_0, \hat{y}_0) .$$

In the sequel we concern ourselves with the problem of the optimal control well-posedness assuming that A3 and the following two conditions hold :

A5. The set V is a compact and convex polyhedron in R^r.The components of the matrices $A_1(t)$ and $A_3(t)$ are in $C^{m-2}(R^1)$ while the components of $A_2(t)$,$A_4^{-1}(t)$,$B_1(t)$ and $B_2(t)$ are in $C^{m-1}(R^1)$.For the matrices $C_j(t)$ defined by the relations

$$C_1(t) = B_0(t) \ , \ C_j(t) = - A_0(t)C_{j-1}(t) + \dot{C}_{j-1}(t), \ j = 2,\dots,m,$$

the general position hypothesis holds, that is if the vector l is parallel to an edge of V, then the vectors $C_1(t)l,\dots,C_m(t)l$ are linearly independent, see Pontrjagin et al. [62] .

A6. The function $g(\cdot)$ is locally Lipschitz continuous.For every solution $(\hat{x}_0,\hat{y}_0)$ of the limit problem (51), if $(p,q) \in \partial_C g(\hat{x}_0,\hat{y}_0)$ then $p - (A_4^{-1}(1)A_3(1))^T q \neq 0$, where $\partial_C g(\cdot)$ is the subgradient defined by Clarke,see Clarke [9].

As it will be further shown, the condition A6 is a sufficient condition for $g_0(x)$ to achieve its minimum at the boundary ∂P of the reachable set P (The general position hypothesis implies that int P is nonempty). Moreover, for sufficiently small β the function $g(x,y)$ achieves its minimum at the boundary of K_β.

From Theorem 3.3 it follows that every L^2- weak limit point of the optimal controls $\hat{u}_\beta(\cdot)$ is an optimal control for the limit problem. We strengthen this result in the following theorem:

<u>Theorem 3.4</u>. For every $\varepsilon > 0$ and for every sequence β_k, $\lim_{k\to\infty} \beta_k = 0$, for which the sequence of optimal controls $\hat{u}_{\beta_k}(\cdot)$ is L^2- weakly convergent to $\hat{u}_o(\cdot)$, there exists $N > 0$ so that if $k > N$ one can choose a finite number of intervals $\Delta_1, \dots, \Delta_p$ such that

$$\text{meas} \bigcup_{i=1}^{p} \Delta_i < \varepsilon$$

and

$$\hat{u}_{\beta_k}(t) = \hat{u}_o(t) \quad \text{for a.e. } t \in [0,1] \setminus \bigcup_{i=1}^{p} \Delta_i .$$

Proof. Let $(\hat{x}_k(\cdot), \hat{y}_k(\cdot))$ be the optimal trajectory corresponding to $\hat{u}_{\beta_k}(\cdot)$ and β_k. We denote by $NM(z_o)$ the normal cone to a convex set $M \subset R^r$ at the point $z_o \in M$, i.e.

$$NM(z_o) = \left\{ l \in R^r , \ l^T(z - z_o) \leqslant 0 \text{ for all } z \in M \right\} .$$

From Theorem 1 in Clarke [9] it follows that for every $k = 1,2,\dots$ there exists a vector

$$(p_k, q_k) \in - \partial_C g(\hat{x}_k(1), \hat{y}_k(1)) \cap NK_{\beta_k}(\hat{x}_k(1), \hat{y}_k(1)) .$$

Moreover, from Lemma 1 in Clarke [9] we conclude that if the sequence $(\hat{x}_k(1), \hat{y}_k(1))$ is convergent, then the sequence (p_k, q_k) possesses a limit point.

Let us assume that the statement of the theorem is false for some $\varepsilon_o > 0$ and for some sequence β_k for which $\lim_{k\to\infty} \hat{u}_{\beta_k}(\cdot) = u_o(\cdot)$ in the weak topology of $L^2(R^r)$. Without loss of generality we suppose that

$$\lim_{k\to\infty} (\hat{x}_k(1), \hat{y}_k(1)) = (\hat{x}_o, \hat{y}_o), \quad \lim_{k\to\infty} (p_k, q_k) = (p_o, q_o) .$$

Clearly, $(\hat{x}_o,\hat{y}_o)$ solves the limit problem (51) and

$$(p_o,q_o) \in NK_o(\hat{x}_o,\hat{y}_o) .$$

We prove that

$$\hat{p} = p_o - (A_4^{-1}(1)A_3(1))^T q_o \in NP_o(\hat{x}_o) . \tag{55}$$

Let $\bar{x} \in P$ and $\bar{u}(\cdot)$ be the corresponding control according to (52). Since $\hat{y}_o \in R(\hat{x}_o)$ there exists $\hat{v}_o(t) \in V$ for $t \in [0,+\infty)$ such that

$$\hat{y}_o = - A_4^{-1}(1)A_3(1)\hat{x}_o + \int_0^\infty \exp(A_4(1)s)B_2(1)\hat{v}_o(s)ds .$$

Define the control

$$\bar{u}_k(t) = \begin{cases} \bar{u}(t) & \text{for } t \in [0,1-\sqrt{\beta_k}] , \\ \hat{v}_o\left(\frac{1-t}{\beta_k}\right) & \text{for } t \in (1-\sqrt{\beta_k}, 1| . \end{cases}$$

If $(\bar{x}_k(\cdot),\bar{y}_k(\cdot))$ corresponds to $\bar{u}_k(\cdot)$ according to (49) for $\beta = \beta_k$, then from Lemma 3.1(ii) it follows that $\lim_{k\to\infty} \bar{x}_k(1) = \bar{x}$. Moreover, from the proof of Theorem 3.1 we get

$$\lim_{k\to\infty} \bar{y}_k(1) = - A_4^{-1}(1)A_3(1)\bar{x} + \int_0^\infty \exp(A_4(1)s)B_2(1)\hat{v}_o(s)ds$$

$$= - A_4^{-1}(1)A_3(1)(\bar{x} - \hat{x}_o) + \hat{y}_o .$$

Then

$$\hat{p}^T(\bar{x} - \hat{x}_o) = p_o^T(\bar{x} - \hat{x}_o) + q_o^T(- A_4^{-1}A_3(1)(\bar{x} - \hat{x}_o))$$

$$= \lim_{k\to\infty} (p_k^T(\bar{x}_k(1) - \hat{x}_k(1)) + q_k^T(\bar{y}_k(1) - \hat{y}_k(1))) \leqslant 0,$$

since $(p_k,q_k) \in NK_{\beta_k}(\hat{x}_k(1),\hat{y}_k(1))$. This proves (55).

Let us denote by $\psi_o(\cdot)$ the solution of the adjoint equation

$$\dot{\psi} = - A_o^T(t)\psi \quad , \psi(1) = \hat{p} . \tag{56}$$

From Proposition 7 in Clarke [9] we have $(p_o,q_o) \in -\partial_C g(x_o,y_o)$. Hence, by A6, $\hat{p} \neq 0$. In Pontrjagin et al. [62] p.202, it is proved that if the general position hypothesis holds and $\psi_o(\cdot) \neq 0$, then there exists a finite number of points $t_2,\ldots,t_{p-1}$ such that $\psi_o^T(t)B_o(t)l = 0$ for every $t \in [\varepsilon_o/4, 1-\varepsilon_o/4] \setminus \{t_2,\ldots,t_{p-1}\}$ and for every vector l which is parallel to an edge of V. Let the intervals $\Delta_1,\ldots,\Delta_p$ be centered at $0,t_2,\ldots,t_{p-1},1$ and let meas $\Delta_j < \varepsilon_o/2p$, $j=1,\ldots,p$. Since $\hat{p} \in NP(\hat{x}_o)$ the optimal control $\hat{u}_o(\cdot)$ is uniquely defined on the set $[0,1] \setminus \bigcup_{j=1}^{p} \Delta_j$ by the Maximum principle

$$\psi_o^T(t)B_o(t)\hat{u}_o(t) = \max_{v \in V} \psi_o^T(t)B_o(t)v \quad . \tag{57}$$

The relations (56) and (57) can be rewritten as

$$\dot{\psi} = - A_1^T(t)\psi - A_3^T(t)\eta \quad , \quad \psi(1) = \hat{p} \ ,$$

$$0 = - A_2^T(t)\psi - A_4^T(t)\eta \quad ,$$

$$(\psi^T(t)B_1(t) + \eta^T(t)B_2(t))\hat{u}_o(t) = \max_{v \in V} (\psi^T(t)B_1(t) + \eta^T(t)B_2(t))v.$$

Since $(p_k,q_k) \in NK_{\beta_k}(\hat{x}_k(1),\hat{y}_k(1))$, then for the perturbed problem we have

$$(\psi_k^T(t)B_1(t) + \eta_k^T(t)B_2(t))\hat{u}_{\beta_k}(t) = \max_{v \in V}(\psi_k^T(t)B_1(t) + \eta_k^T(t)B_2(t))v,$$

where $(\psi_k(\cdot),\eta_k(\cdot))$ solves

$$\dot{\psi} = - A_1^T(t)\psi - A_3^T(t)\eta \quad , \quad \psi(1) = p_k \ ,$$

$$\beta_k\dot{\eta} = - A_2^T(t)\psi - A_4^T(t)\eta \quad , \quad \eta(1) = q_k/\beta_k \ .$$

From Lemma 3.1(i) and (v) it follows that

$$\lim_{k\to} \max_{0 \leq t \leq 1-\varepsilon_o/2} (|\psi_k(t) - \psi(t)| + |\eta_k(t) - \eta(t)|) = 0.$$

Hence, for sufficiently large k and for every vector l, which is parallel to an edge of V we have

$$\xi_k^T(t)l = (\psi_k^T(t)B_1(t) + \eta_k^T(t)B_2(t))l \neq 0 ,$$

and

$$\max_{v \in V} \xi_k^T(t)v = \xi_k^T(t)\hat{u}_o(t)$$

for every $t \in [0,1] \setminus \bigcup_{j=1}^{p} \Delta_j$. The obtained contradiction completes the proof.

Remark 3.2. If the limit problem (51) has an unique solution, then Theorem 3.4 can be formulated in the following way: for every $\varepsilon > 0$ there exists $\Lambda > 0$ such that if $\beta \in (0,\Lambda)$ and $\hat{u}_\beta(\cdot)$ is an optimal control for (48) then there exists a finite number of intervals $\Delta_1, \ldots, \Delta_p$ such that meas $\bigcup_{j=1}^{p} \Delta_j < \varepsilon$ and $\hat{u}_\beta(t) = \hat{u}_o(t)$ for a. e. $t \in [0,1] \setminus \bigcup_{j=1}^{p} \Delta_j$.

In order to obtain uniqueness of $\hat{u}_o(\cdot)$ it is sufficient to assume that $g(\cdot)$ is convex. Let us prove this statement. Let $(\hat{x}_1, \hat{y}_1) \in K_o$ and $(\hat{x}_2, \hat{y}_2)$ be two different solutions of (51). Then

$$(\hat{x}_o, \hat{y}_o) = ((\hat{x}_1, \hat{y}_1) + (\hat{x}_2, \hat{y}_2))/2$$

is a solution. Moreover, since the set P is strictly convex, we have $\hat{x}_o \in$ int P. From Theorem 1 in Clarke [9] we deduce that there exists $(\tilde{p}, \tilde{q}) \in -\partial_C g(\hat{x}_o, \hat{y}_o) \cap NK_o(\hat{x}_o, \hat{y}_o)$. Let $x \in P$ be arbitrarily chosen and let $y = -A_4^{-1}(1)A_3(1)(x - \hat{x}_o) + \hat{y}_o$. Clearly, $y \in R(x)$, and

$$0 \geq \tilde{p}^T(x - \hat{x}_o) + \tilde{q}^T(y - \hat{y}_o) = (\tilde{p} - (A_4^{-1}(1)A_3(1))^T\tilde{q})(x - \hat{x}_o),$$

which, combined with $\hat{x}_o \in$ int P implies that $\tilde{p} - (A_4^{-1}(1)A_3(1))^T\tilde{q} = 0$. This contradicts assumption A6. Thus, the general position hypothesis implies that the optimal control $\hat{u}_o(\cdot)$ is unique, see Pontrjagin et al. [62], p.139.

By repeating the arguments of Theorem 3.2 in Gičev and Dontchev [30] one can prove

Theorem 3.5. Suppose that the optimal control $\hat{u}_o(\cdot)$ is unique. Then for every $\varepsilon > 0$ there exists $\Lambda > 0$ such that if $\beta \in (0,\Lambda)$ and $(\hat{x}_\beta(\cdot), \hat{y}_\beta(\cdot))$ is an optimal trajectory for the perturbed problem

(48) , then there exists a finite number of intervals $\Delta_1,\dots,\Delta_p$ such that meas $\bigcup_{j=1}^{p} \Delta_j < \varepsilon$ and

$$\max_{t\in[0,1]} |\hat{x}_\beta(t) - \hat{x}_o(t)| + \sup_{t\in[0,1]\setminus\bigcup_{j=1}^{p}\Delta_j} |\hat{y}_\beta(t) - \hat{y}_o(t)| < \varepsilon ,$$

where $\hat{x}_o(\cdot)$ is the optimal trajectory for the reduced system and $\hat{y}_o(t) = - A_4^{-1}(t)(A_3(t)\hat{x}_o(t) + B_2(t)\hat{u}_o(t))$, $t\in[0,1]$.

3.4.2. Lagrange problem

The next classical problem in optimal control we shall study in presence of singular perturbations is to minimize the functional

$$I(x(\cdot),y(\cdot),u(\cdot)) = \int_0^1 f(x(t),y(t),u(t),t)dt \tag{58}$$

subject to

$$\begin{aligned} \dot{x} &= A_1(t)x + A_2(t)y + B_1(t)u , \\ \beta\dot{y} &= A_3(t)x + A_4(t)y + B_2(t)u , \end{aligned} \tag{59}$$

$$x(0) = x^o , \; x(1) = x^1 , \tag{60a}$$

$$y(0) = y^o , \; y(1) = y^1 , \tag{60b}$$

$$u(\cdot) \in U = \{u(\cdot)\in L^1(R^r),\ u(t)\in V \text{ for a.e. } t\in[0,1]\}. \tag{61}$$

For $\beta = 0$ we get the reduced system

$$\dot{x} = A_o(t)x + B_o(t)u , \tag{62a}$$

$$y(t) = - A_4^{-1}(t)(A_3(t)x(t) + B_2(t)u(t)),\ t\in[0,1] , \tag{62b}$$

and the corresponding limit problem consists in minimizing (58) subject to (60a) ,(61) and (62a,b).

The main result in this paragraph is obtained under the condition A3(concerning the continuity of the matrices and the spectrum of $A_4(t)$), and the following conditions:

A7.(i) The set V is compact and convex.

(ii) The function $f(\cdot)$ satisfies the Caratheodory condition, i.e. it is continuous with respect to (x,y,u) and measurable with respect to t in appropriate sets.The functional $I(\cdot)$ is lower semicontinuous in the uniform topology for $x(\cdot)$ and in the L^2-weak topology for $(y(\cdot),u(\cdot))$.

(iii) As before,denote by P the reachable set at t=1 of the reduced system (62a) and let

$$R = \int_0^{\infty} \exp(A_4(1)s)B_2(1)V ds \ .$$

It is assumed that $x^1 \in \text{int } P$ and $y^1 + A_4^{-1}(1)A_3(1)x^1 \in \text{int } R$.

Notice that the sets P and R have nonempty interiors if, for example, int $V \neq \emptyset$,the system (62a) is controllable with unrestricted controls and the pair $(A_4(1),B_2(1))$ satisifes Kalman's rank condition.

By a standard argument there exists an optimal control $\hat{u}_0(\cdot)$ for the reduced problem.Moreover, from Theorem 3.1 we conclude that for small β the target point (x^1,y^1) belongs to the reachable set K_β of the system (59), hence there exists an optimal control $\hat{u}_\beta(\cdot)$ for the perturbed problem (58).In the sequel it is assumed that β is sufficiently small.

Denote by $\hat{x}_0(\cdot)$ the optimal trajectory and by $\hat{I}_0$ the optimal value of the limit problem.Let $\hat{y}_0(\cdot)$ be determined by (62b) for $\hat{x}_0(\cdot)$ and $\hat{u}_0(\cdot)$.The optimal solution of the perturbed problem is denoted by $(\hat{x}_\beta(\cdot),\hat{y}_\beta(\cdot),\hat{u}_\beta(\cdot))$ and $\hat{I}_\beta$.

Theorem 3.6. The problem (58) is well-posed in the sense of performance convergence, that is

$$\lim_{\beta \to 0} \hat{I}_\beta = \hat{I}_0 \ .$$

Proof.We use the general scheme from § 2.2.1.Let the sequence β_k , $\lim_{k\to\infty} \beta_k = 0$,be arbitrarily chosen.Recall that the Hausdorff limit of the reachable set K_β is

$$K_0 = \left\{ (x,y) \in R^{m+n} \ , \ x \in P \ , \ y \in - A_4^{-1}(1)A_3(1)x + R \right\} .$$

From A7(iii) it follows that for $\varepsilon > 0$ sufficiently small there exists a simplex $G = \left\{(x_\varepsilon^1,y_\varepsilon^1),\dots,(x_\varepsilon^p,y_\varepsilon^p)\right\}$, $p = m+n+1$, such that $G \subset K_0$, $|x_\varepsilon^i - x^1| < \varepsilon$,$i = 1,\dots,p$, and for some $d > 0$

$$\left\{ (x,y)\in R^{m+n} , |x - x^1| < d\varepsilon , |y - y^1| < d\varepsilon \right\} \subset G .$$

Using Lemma 3.3 we conclude that there exist controls $u^i_\varepsilon(\cdot) \in U$, $i=1,\dots,p$, such that the corresponding states of (62a) satisfy $x(1) = x^i_\varepsilon$, $i= 1,\dots,p$, and $u^i_\varepsilon(t) = \hat{u}_o(t)$ for a.e. $t \in [0,1] \setminus T^i(\varepsilon)$, where meas $T^i(\varepsilon) \leqslant c_o\varepsilon$, $i=1,\dots,p$. On the other hand there exist functions $v^i_\varepsilon(\cdot)$, $v^i(t) \in V$ for a.e. $t\in[0,+\infty)$, $i=1,\dots,p$, such that

$$y^i = - A_4^{-1}(1)A_3(1)x^i + \int_0^\infty \exp(A_4(1)s)B_2(1)v^i_\varepsilon(s)ds$$

for $i=1,\dots,p$. Let the sequence φ_k satisfy $\lim_{k\to\infty} \varphi_k = 0$, $\lim_{k\to\infty} \varphi_k/\beta_k = +\infty$, $\varphi_k > 0$, $k=1,2,\dots$ Define the control

$$\tilde{u}^i_k(t) = \begin{cases} u^i_\varepsilon(t) & \text{for } t\in [0,1-\varphi_k) , \\ v^i_\varepsilon\left(\frac{1-t}{\beta_k}\right) & \text{for } t\in [1-\varphi_k,1] , \end{cases}$$

and let $(\tilde{x}^i_k(\cdot),\tilde{y}^i_k(\cdot))$ be the corresponding trajectory of (59) for β_k. As in the proof of Theorem 3.1 we get

$$\lim_{k\to\infty} \tilde{x}^i_k(1) = x^i_\varepsilon \quad , \qquad \lim_{k\to\infty} \tilde{y}^i_k(1) = y^i_\varepsilon .$$

Then for k sufficiently large

$$(x^1,y^1) \in co((\tilde{x}^i_k(1),\tilde{y}^i_k(1)))_i .$$

Hence there exist numbers $\alpha^1_k,\dots, \alpha^p_k \geqslant 0$, the sum of which is 1, such that if

$$\bar{u}^\varepsilon_k(\cdot) = \sum_{i=1}^{p} \alpha^i_k \tilde{u}^i_k(\cdot)$$

and $(\bar{x}^\varepsilon_k(\cdot),\bar{y}^\varepsilon_k(\cdot))$ corresponds to $\bar{u}^\varepsilon_k(\cdot)$ and β_k according to (59), then for k sufficiently large

$$\bar{x}_k(1) = x^1 \quad \text{and} \quad \bar{y}_k(1) = y^1 .$$

Without loss of generality, let $\lim_{k\to\infty} \alpha^i_k = \alpha^i$. Then $\sum_{i=1}^{p} \alpha^i = 1$.

Define

$$\bar{u}_o^{\varepsilon}(\cdot) = \sum_{i=1}^{p} \alpha^i u_{\varepsilon}^i(\cdot) .$$

It is clear that $\bar{u}_o^{\varepsilon}(t) = \hat{u}_o(t)$ for a.e. $t \in [0,1] \setminus \bigcup_{i=1}^{p} T^i(\varepsilon)$ and $\bar{u}_k^{\varepsilon}(\cdot)$ is pointwise convergent to $\bar{u}_o^{\varepsilon}(\cdot)$.Let $(\bar{x}_o^{\varepsilon}(\cdot),\bar{y}_o^{\varepsilon}(\cdot))$ correspond to $\bar{u}_o^{\varepsilon}(\cdot)$ according to (62).Applying Lemma 3.1 (iv) we get that $\bar{x}_k^{\varepsilon}(\cdot)$ converges to $\bar{x}_o^{\varepsilon}(\cdot)$ uniformly in $[0,1]$ and $\bar{y}_k^{\varepsilon}(\cdot)$ is L^2- strongly convergent to $\bar{y}_o^{\varepsilon}(\cdot)$.Thus,one can consider $\bar{y}_k^{\varepsilon}(\cdot)$ as a pointwise convergent sequence.Since $\bar{u}_k^{\varepsilon}(\cdot)$ and $\bar{y}_k^{\varepsilon}(\cdot)$ are uniformly bounded,we obtain

$$\limsup_{k\to\infty} \hat{I}_{\beta_k} \leqslant \lim_{k\to\infty} I(\bar{x}_k^{\varepsilon}(\cdot),\bar{y}_k^{\varepsilon}(\cdot),\bar{u}_k^{\varepsilon}(\cdot)) = I(\bar{x}_o^{\varepsilon}(\cdot),\bar{y}_o^{\varepsilon}(\cdot),\bar{u}_o^{\varepsilon}(\cdot)). \quad (63)$$

Now choose a sequence ε_i ,$\lim_{i\to\infty} \varepsilon_i = 0$.Then $\bar{u}_o^{\varepsilon_i}(\cdot)$ is convergent in measure to $\hat{u}_o(\cdot)$,hence it can be identified with a pointwise convergent sequence.Clearly, the trajectory $\bar{x}_o^{\varepsilon_i}(\cdot)$ is uniformly convergent to $\hat{x}_o(\cdot)$ in $[0,1]$ and $y_o^{\varepsilon_i}(\cdot)$ is pointwise convergent to $\hat{y}_o(\cdot)$.Hence

$$\lim_{i\to\infty} I(x_o^{\varepsilon_i}(\cdot),y_o^{\varepsilon_i}(\cdot),u_o^{\varepsilon_i}(\cdot)) = \hat{I}_o . \quad (64)$$

On the other hand,the sequence of optimal controls $\hat{u}_{\beta_k}(\cdot)$ can be thought of as a L^2-weakly convergent sequence to some $u_o(\cdot) \in U$.From Lemma 3.1(ii) we get that the optimal trajectory $\hat{x}_{\beta_k}(\cdot)$ is uniformly convergent to $x_o(\cdot)$ which satisfies (60a) and (62a) for $u_o(\cdot)$, and $\hat{y}_{\beta_k}(\cdot)$ is L^2-weakly convergent to $y_o(\cdot)$,which satisfies (62b) for $x_o(\cdot)$ and $u_o(\cdot)$.Hence,by A7(ii)

$$\hat{I}_o \leqslant I(x_o(\cdot),y_o(\cdot),u_o(\cdot)) \leqslant \liminf_{k\to\infty} \hat{I}_{\beta_k} . \quad (65)$$

Combining (63) and (64) and taking into account (65) we conclude that from every sequence β_k one can extract a subsequence β_l such that

$$\lim_{l\to\infty} \hat{I}_{\beta_l} = \hat{I}_o \quad , \text{Q.E.D.}$$

3.4.3. Generalizations

I) State constraints. We show that Theorem 3.1 can be extended to problems with additional state constraints of the form

$$x(t) \in X \quad \text{for all } t \in [0,1] \quad , \tag{66}$$

assuming that A3 and the following condition hold:

A8. The sets $V \subset R^r$ and $X \subset R^m$ are closed and convex. There exists a feasible control $\tilde{u}(\cdot)$ such that the corresponding solution $\tilde{x}(\cdot)$ of the reduced system (62a) satisfies

$$\tilde{x}(t) \in \operatorname{int} X \quad \text{for all } t \in [0,1] .$$

Denote by U_X the admissible set of controls for the reduced system with (66), that is: $u(\cdot) \in U$ and the corresponding state of (62a) satisfies (66). As before, let P be the reachable set of (62a) with controls from U_X, and let K_β be the reachable set for the full-order system with (66) (at t=1). The sets R(x) and K_o are defined as in Section 3.3 .

One can modify the proof of Theorem 3.1 in the following way:

Let $(x_o, y_o) \in K_o$, $u_o(\cdot) \in U_X$, $v_o(t) \in V$ for $t \in [0,+\infty)$ and $u_\beta(\cdot)$ be chosen as in the proof of Theorem 3.1. By Lemma 3.1(ii) we get that $\lim_{\beta \to 0} \|x_\beta - x_o\|_C = 0$, where $x_\beta(\cdot)$ solves (59) for $u_\beta(\cdot)$. There exists a function $\varepsilon(\beta)$ with values in (0,1) such that

$$\lim_{\beta \to 0} \varepsilon(\beta) = 0 \quad \text{and} \quad \lim_{\beta \to 0} (\|x_\beta - x_o\|_C + \sqrt{\beta})/\varepsilon(\beta) = 0 .$$

Define the control

$$\bar{u}_\beta(t) = \begin{cases} (1-\varepsilon(\beta))u_o(t) + \varepsilon(\beta)\tilde{u}(t) & \text{for } t \in [0,1-\sqrt{\beta}] , \\ v_o((1-t)/\beta) & \text{for } t \in (1-\sqrt{\beta}, 1] . \end{cases}$$

Clearly, $\bar{u}_\beta(\cdot) \in U$. Let $(\bar{x}_\beta(\cdot), \bar{y}_\beta(\cdot))$ correspond to $\bar{u}_\beta(\cdot)$ from (59). Since $x_o(t) \in X$ one can easily deduce that

$$\bar{x}_\beta(t) = (1-\varepsilon(\beta))x_\beta(t) + \varepsilon(\beta)\tilde{x}_\beta(t) \in X \quad \text{for all } t \in [0,1-\sqrt{\beta}] .$$

We prove that there exists a constant $\alpha > 0$ such that

$$\operatorname{dist}(\bar{x}_\beta(1-\sqrt{\beta}),\partial X) \geq \alpha\varepsilon(\beta) .$$

Denote $t_\beta = 1-\sqrt{\beta}$ and let

$$|\bar{x}_\beta(t_\beta) - z_\beta| = \operatorname{dist}(\bar{x}_\beta(t_\beta),\partial X) ,$$

$$l_\beta = z_\beta - (1-\varepsilon(\beta))x_o(t_\beta) - \varepsilon(\beta)\tilde{x}(t_\beta) .$$

Since $z_\beta \in \partial X$ and $\tilde{x}(t_\beta) \in \operatorname{int} X$, then $l_\beta \neq 0$.From A8 it follows that there exists $\delta > 0$ such that $\tilde{x}(t_\beta) + \delta l_\beta/|l_\beta| \in X$.Then

$$(1-\varepsilon(\beta))x_o(t_\beta) + \varepsilon(\beta)\tilde{x}(t_\beta) + \delta\varepsilon(\beta)l_\beta/|l_\beta| \in X .$$

By the definition of l_β we get $\delta\varepsilon(\beta) \leq |l_\beta|$.For β sufficiently small we have

$$\begin{aligned}|\bar{x}_\beta(t_\beta) - z_\beta| &= |(1-\varepsilon(\beta)x_\beta(t_\beta) + \varepsilon(\beta)\tilde{x}_\beta(t_\beta) - z_\beta| \\ &\geq |l_\beta| - (1-\varepsilon(\beta))|x_\beta(t_\beta) - x_o(t_\beta)| \\ &\quad - \varepsilon(\beta)|\tilde{x}_\beta(t_\beta) - \tilde{x}(t_\beta)| \\ &\geq \delta\varepsilon(\beta) - (1-\varepsilon(\beta))\|x_\beta - x_o\|_C - \varepsilon(\beta)\|\tilde{x}_\beta - \tilde{x}\|_C \\ &\geq \delta\varepsilon(\beta)/2 .\end{aligned}$$

Thus , since

$$|\bar{x}_\beta(t_\beta) - \bar{x}_\beta(t)| = O(\sqrt{\beta}) \quad \text{for all } t \in [t_\beta,1] ,$$

and $\lim_{\beta\to 0} \sqrt{\beta}/\varepsilon(\beta) = 0$, for β sufficiently small we obtain that $\bar{x}_\beta(t) \in X$ for all $t \in [0,1]$ .This means that $\bar{u}_\beta(\cdot) \in U_x$.One can also prove that $\lim_{\beta\to 0} \bar{x}_\beta(1) = x_o$ and $\lim_{\beta\to 0} \bar{y}_\beta(1) = y_o$.The further proof is completely analogous to the proof of Theorem 3.1.

The presence of state constraints for the fast variables complicates considerably the situation.The following example shows that even in the case when the function $g(\cdot)$ (the terminal part) in the performance index is independent of the fast states, the substitution $\beta = 0$ in the perturbed system does not define a limit problem.

Example 3.2. Minimize $x^2(1)$ subject to

$$\begin{aligned} \dot{x} &= y_1 & x(0) &= -1 , \\ \beta\dot{y}_1 &= - y_1 + y_2 & y_1(0) &= 0 , \\ \beta\dot{y}_2 &= \quad - y_2 + u , & y_2(0) &= e , \end{aligned}$$

$$u(t) \in [0,1] , \ y_1(t) \in [-1,1] , \ t \in [0,1] .$$

We have

$$y_2(t) = e^{-t/\beta} e + \frac{1}{\beta}\int_0^t e^{-(t-s)/\beta} u(s) ds \geqslant e^{1-t/\beta} ,$$

$$y_1(t) = \frac{1}{\beta}\int_0^t e^{-(t-s)/\beta} y_2(s) ds \geqslant \frac{t}{\beta} e^{1-t/\beta} .$$

The only feasible control is $u(t) = 0$, which gives value 1 for the objective function.Note that the reachable set for the slow state x consists of one point.

The "reduced" problem,obtained for $\beta = 0$ is

$$\begin{aligned} & x^2(1) \longrightarrow \inf \\ & \dot{x} = u , \ x(0) = -1 , \\ & u(t) \in [0,1] , \end{aligned}$$

which solution $\hat{u}(t) = 1$ gives $x(1) = 0$.

2) The problem of Bolza. Consider the problem

$$J_\beta(u(\cdot)) = g(x(1),y(1)) + \int_0^1 f(x(t),y(t),u(t),t)dt \longrightarrow \inf$$

subject to (59) and (61), where the function $g(\cdot)$ is continuous, the set V is compact and convex, and the integral part of $J_\beta(\cdot)$ satisfies the conditions in A7(ii).The objective function for the limit problem will have the form

$$J_o(u(\cdot)) = g_o(x(1)) + \int_0^1 f(x(t),-A_4^{-1}(t)(A_3(t)x(t) + B_2(t)u(t)),u(t),t)dt ,$$

where $g_o(\cdot)$ is defined in (54) and the state $x(\cdot)$ corresponds to $u(\cdot)$ according to (62a).

On these assumptions, the problem in question is well-posed in the sense of performance convergence.We briefly outline the proof, which uses the general scheme from § 2.2.1.

Choosing a L^2 - weakly convergent subsequence of the optimal controls $\hat{u}_\beta(\cdot)$ and applying Lemma 3.1 one can get

$$J_o(\hat{u}_o(\cdot)) \leqslant \liminf_{\beta \to 0} J_\beta(\hat{u}_\beta(\cdot)) \quad ,$$

where $\hat{u}_o(\cdot)$ is an optimal control for the limit problem.

Let $\hat{x}_o(\cdot)$ be an optimal trajectory for the limit problem and $\hat{y}_o$ satisfy $g_o(\hat{x}_o(1)) = g(\hat{x}_o(1),\hat{y}_o)$. Then there exists a function $v(\cdot)$, $v(t) \in V$ for a.e.$t \in [0,+\infty)$ such that

$$\hat{y}_o = - A_4^{-1}(1)A_3(1)\hat{x}_o(1) + \int_0^\infty \exp(A_4(1)s)B_2(1)v(s)ds \ .$$

Define the control

$$u_\beta(t) = \begin{cases} \hat{u}_o(t) & \text{for } t \in [0,1-\sqrt{\beta}] \ , \\ v(\frac{1-t}{\beta}) & \text{for } t \in (1-\sqrt{\beta},1] \ . \end{cases}$$

From Lemma 3.1(iv) the corresponding trajectory $(x_\beta(\cdot),y_\beta(\cdot))$ of the perturbed system converges L^2 - strongly to the optimal trajectory $(\hat{x}_o(\cdot),\hat{y}_o(\cdot))$ of the limit problem.Hence,it can be thought of as a pointwise convergent sequence.Finally we obtain

$$\limsup_{\beta \to 0} J_\beta(\hat{u}_\beta(\cdot)) \leqslant \limsup_{\beta \to 0} J_\beta(u_\beta(\cdot)) = J_o(\hat{u}_o(\cdot)),$$

which gives us performance well-posedness.

3) Unrestricted controls. As it was noted in § 2.2.1, under some grouth condition for the functional,our scheme can be applied to problems with unbounded admissible sets of controls.

Let us consider the problem: minimize the functional

$$J_\beta(u(\cdot)) = g(x(1),y(1)) + \int_0^1 (f(x(t),y(t),t) + h(u(t),t))dt$$

over the space $L^2(R^r)$, where $(x(\cdot),y(\cdot))$ is determined by the singularly perturbed system (59).

Step 1: Definition of the limit problem.

Suppose that the function $g(\cdot)$ is continuous and there exists a continuous function $Q: R^m \to R^n$ such that

$$g(x,Q(x)) \leqslant g(x,y)$$

for all $x \in R^m$ and $y \in R^n$. We define the limit functional as

$$J_0^*(u(\cdot)) = g(x(1),Q(x(1)) + \int_0^1 (f(x(t),y(t),t) + h(u(t),t))dt ,$$

where $(x(\cdot),y(\cdot))$ is determined by (62ab).

If the function $g(\cdot)$ is not dependent of x, the above condition means that $g(\cdot)$ is bounded below over R^n. The following example shows that such a condition is essential for performance convergence.

Example 3.3.

$$J_\beta(u(\cdot)) = c^T y(1) + \int_0^1 |u(t)|^2 dt \quad , c \neq 0 ,$$

$$\beta \dot{y} = A_4 y + u \quad , y(0) = y^0 .$$

From Theorem 3.2 it follows that the reachable set of the system tends to the entire space when $\beta \to 0$. Then for every $a > 0$ and for every sequence $\beta_k, \lim_{k\to\infty} \beta_k = 0$, there exists an uniformly bounded sequence of controls $u_k(\cdot)$, pointwise convergent to zero, such that the corresponding final state satisfies

$$\lim_{k\to\infty} y_k(1) = -ac .$$

Hence

$$\lim_{k\to\infty} J_{\beta_k}(\hat{u}_{\beta_k}(\cdot)) \leqslant \lim_{k\to\infty} J_{\beta_k}(u_k(\cdot)) = -a|c|^2 .$$

Since a is arbitrarily chosen

$$\lim_{\beta\to 0} J_\beta(\hat{u}_\beta(\cdot)) = -\infty .$$

Step 2: Lower bound.

Let $g(\cdot)$ and $f(\cdot)$ be bounded below and $h(\cdot)$ satisfy

$$h(u,t) \geq b|u|^2 , \quad t\in[0,1]$$

for some $b>0$ and for all $u\in R^r$. It is easy to show that in this case every sequence of optimal controls $\hat{u}_\beta(\cdot)$ is bounded in $L^2(R^r)$ when $\beta\to 0$. Assuming that the integral part of the functional is L^2 - weakly lower semicontinuous we get

$$J_0^*(\hat{u}_0(\cdot)) \leq \liminf_{\beta\to 0} J_\beta^*(\hat{u}_\beta(\cdot)) \leq \liminf_{\beta\to 0} J_\beta(\hat{u}_\beta(\cdot)) .$$

Step 3: Upper bound.

We have to define a feasible control $u_\beta(\cdot)$ for the perturbed system such that

$$\lim_{\beta\to 0} J_\beta(u_\beta(\cdot)) = J_0^*(\hat{u}_0(\cdot)).$$

If

$$Q(\hat{x}_0(1)) \in - A_4^{-1}(1)A_3(1)\hat{x}_0(1) + A_4(1)/B_2(1)$$

one can apply Theorem 3.2 choosing a sequence of controls $u_\beta(\cdot)$ pointwise convergent to $\hat{u}_0(\cdot)$ such that the corresponding trajectory $(x_\beta(\cdot),y_\beta(\cdot))$ satisfies

$$\lim_{\beta\to 0} (\|x_\beta - \hat{x}_0\|_C + |y_\beta(1) - Q(\hat{x}_0(1))| + \|y_\beta - \hat{y}_0\|) = 0.$$

Hence, imposing some natural restrictions for $f(\cdot)$ and $g(\cdot)$ we get performance convergence.

In contrast to the bounded case, problems with unrestricted controls have been investigated recently in a number of papers. We mention here Dragan and Halanay [21] , Glizer and Dmitriev [32] and O'Malley [57] . In the author's opinion, the interest in such problems comes from the fact that, as a rule, these problems can be reduced to systems of singularly perturbed differential equations. Cleary, when the functional is quadratic with respect to the control, applying the Maximum principle, we get a singularly perturbed adjoint equation, to which, together with the system equation, an asymptotic series analysis can be used. A technique, similar to that presented here, was used in Dontchev and Gičev [18] , [28] and Dontchev [17] to strictly convex problems.

3.5. Estimations

3.5.1. Lagrange problem (continuation)

In this paragraph we develop the analysis from § 3.4.2 estimating the performance convergence rate for the singularly perturbed Lagrange problem

$$I(x(\cdot),y(\cdot),u(\cdot)) = \int_0^1 f(x(t),y(t),u(t),t)dt \longrightarrow \inf \qquad (67)$$

subject to (59) through (61), assuming that the conditions A3 and A7 hold and

A9. The components of the matrices $A_2(t), A_3(t), A_4(t)$ and $B_2(t)$ are in C^1. The function $f(\cdot)$ is Lipschitz continuous with respect to (x,y) for bounded (x,y) uniformly in $(u,t) \in V \times [0,1]$.

The limit problem for the reduced system (62) is defined as in § 3.4.2. We denote by $\hat{I}_B$ and $\hat{I}_o$ the optimal values and by $\hat{u}_B(\cdot)$ and $\hat{u}_o(\cdot)$ the optimal controls of the perturbed and the limit problems respectively.

Theorem 3.7. There exists a constant c such that

$$-c(B + \omega(\hat{u}_B,B)_1) \leqslant \hat{I}_B - \hat{I}_o \leqslant c(B + \omega(\hat{u}_o,B)_1) .$$

Proof. Although in general outline the proof goes parallelly to the proof of Theorem 3.6, it differs from that in several important details. Therefore, we present here the complete proof.

Choose vectors $\xi_i \in R^m$, $|\xi_i| = 1$, $i = 0,\ldots,m$, such that $0 \in \text{int co}(\xi_i)_i$. There exists $\alpha > 0$ such that for every $\bar{\xi}_i \in R^m, i = 0, \ldots,m, |\bar{\xi}_i - \xi_i| \leqslant \alpha$ one has $0 \in \text{co}(\bar{\xi}_i)_i$. Let $x_i^\varepsilon = x^1 + \varepsilon\xi_i, i = 0,\ldots,m$, where $\varepsilon > 0$. From the choise of ξ_i is follows that if $|\bar{x}_i^\varepsilon - x_i^\varepsilon| < \alpha\varepsilon$ then $x^1 \in \text{co}(\bar{x}_i^\varepsilon)_i$. Analogically, let $\eta_j \in R^n, |\eta_j| = 1, j = 0,\ldots,n$, $0 \in \text{int co}(\eta_j)_j$, and $y_j = y^1 + d\eta_j$. For sufficiently small ε and d we have $(x_i^\varepsilon, y_j) \in \text{int } K_o$. There exists $\varkappa > 0$ such that if $|\bar{\eta}_j - \eta_j| \leqslant \varkappa$ $j = 0,\ldots,n$, then $0 \in \text{co}(\bar{\eta}_j)_j$. Thus, if $|\bar{y}_j - y_j| \leqslant \varkappa d$, then $y^1 \in \text{co}(\bar{y}_j)_j$

Using Lemma 3.3, for sufficiently small ε we choose a control $u_i^\varepsilon(\cdot)$ which drives the state of (62a) from x^o at t=0 to x_i^ε at t = 1 and which differs from $\hat{u}_o(\cdot)$ on a set of measure c_o. Since

$$y_j \in - A_4^{-1}(1)A_3(1)x_i^\varepsilon \quad + R, \ i=0,\dots,m, \ j=0,\dots,n,$$

there exists a measurable function $v_{ij}^\varepsilon(\cdot), v_{ij}^\varepsilon(t) \in V$ for $t \in [0,+\infty)$ such that

$$y_j = - A_4^{-1}(1)A_3(1)x_i^\varepsilon + \int_0^\infty \exp(A_4(1)s)B_2(1)v_{ij}^\varepsilon(s)ds \ .$$

For small β define the control

$$\tilde{u}_{ij}^\beta(t) = \begin{cases} u_i^\varepsilon(t) & \text{for } t \in [0,1-\theta\beta] \ , \\ v_{ij}^\varepsilon(\frac{1-t}{\beta}) & \text{for } t \in (1-\theta\beta,1] \ , \end{cases}$$

where $\theta \geqslant 1$ is arbitrarily chosen.If $(\tilde{x}_{ij}^\beta(\cdot), \tilde{y}_{ij}^\beta(\cdot))$ is the corresponding trajectory of (59),applying Lemma 3.2 we have

$$|\tilde{x}_{ij}^\beta(1) - x_i^\varepsilon| \leqslant c_1(\theta + 1)\beta \ ,$$

where c_1 does not depend on ε , θ and β.Using this estimate we get

$$\frac{1}{\beta}\int_0^1 \exp(A_4(1)\ \frac{1-t}{\beta})A_3(1)\tilde{x}_{ij}^\beta(t)dt = - A_4^{-1}(1)A_3(1)\tilde{x}_{ij}^\beta(1)$$

$$+ A_4^{-1}(1)\ \exp(A_4(1)/\beta)A_3(1)x^0 - \frac{1}{\beta}\int_0^1\int_0^t \exp(A_4(1)\ \frac{1-s}{\beta})ds \ \times$$

$$A_3(1)\tilde{x}_{ij}^\beta(t)ds = - A_4^{-1}(1)A_3(1)x_i^\varepsilon \quad + h(\theta,\beta) \ , \tag{68}$$

where $h(\theta,\beta) \leqslant c_4(\theta + 1)\beta$.Denote by $\bar{y}_{ij}^\beta(\cdot)$ the solution of the equation

$$\beta\dot{y} = A_3(1)\tilde{x}_{ij}^\beta(t) + A_4(1)y + B_2(1)\tilde{u}_{ij}^\beta(t), \ y(0) = y^0 \ .$$

As in Theorem 3.1 one can prove that $|\tilde{y}_{ij}^\beta(1) - \bar{y}_{ij}^\beta(1)| \leqslant \delta_\beta$, where δ_β is not dependent on $\tilde{u}_{ij}^\beta(\cdot)$ and $\lim_{\beta \to 0} \delta_\beta = 0$.Using (68) we have

$$|\tilde{y}_{ij}^\beta(1) - y_j| \leqslant \delta_\beta + |\bar{y}_{ij}^\beta(1) - y_j| \leqslant \delta_\beta + \sigma_0\exp(-\sigma/\beta)\,|y^0|$$

$$+ \left|-y_j - A_4^{-1}(1)A_3(1)x_i^\varepsilon + h(\theta,\beta) + \frac{1}{\beta}\int_{1-\theta\beta}^{1} \exp(A_4(1)\frac{1-t}{\beta})B_2(1) \times\right.$$

$$\left. v_{ij}^\varepsilon(\frac{1-t}{\beta})dt \right| + \left|\frac{1}{\beta}\int_0^{1-\theta\beta} \exp(A_4(1)\frac{1-t}{\beta})B_2(1)u_i^\varepsilon(t)dt\right|$$

$$\leqslant \delta_\beta + c_5(\beta + \theta\beta + e^{-\sigma\theta}) \leqslant \gamma(\theta) + \theta\varphi(\beta) ,$$

where $\lim\limits_{\theta\to\infty}\gamma(\theta) = 0, \lim\limits_{\beta\to 0}\varphi(\beta) = 0$, $\gamma(\cdot)$ is not dependent of ε and β and $\varphi(\cdot)$ is not dependent of ε and θ .Fix θ such that $\gamma(\theta) < \varkappa d/2$. Then for every ε and β satisfying

$$c_1(\theta + 1)\beta < \alpha\varepsilon \quad \text{and} \quad \varphi(\beta) < \varkappa d/2\theta, \tag{69}$$

we have

$$x^1 \in co(\tilde{x}_{ij}^\beta(1))_i , \; j = 0,\dots,n.$$

This means that there exist $\alpha_{ij} \geqslant 0$, $\sum\limits_{i=0}^{m}\alpha_{ij} = 1$ such that

$$x^1 = \sum_{i=0}^{m}\alpha_{ij}\tilde{x}_{ij}^\beta(1) .$$

If

$$\bar{\bar{y}}_j = \sum_{i=0}^{m}\alpha_{ij}\tilde{y}_{ij}^\beta(1),$$

then

$$|\bar{\bar{y}}_j - y_j| \leqslant \varkappa d ,$$

hence one can choose $\delta_j \geqslant 0$, $\sum\limits_{j=0}^{n}\delta_j = 1$, so that $y^1 = \sum\limits_{j=0}^{n}\delta_j y_j$. Finally, we get that

$$(x^1,y^1) = \sum_{i=0}^{m}\sum_{j=0}^{n} \alpha_{ij}\delta_j(\tilde{x}_{ij}^\beta(1),\tilde{y}_{ij}^\beta(1)).$$

Then there exists a control $u_\beta^\varepsilon(\cdot)$, which is a convex combination of $\tilde{u}_{ij}^\beta(\cdot)$, and which drives the state of (59) to (x^1,y^1).Moreover, $u_\beta^\varepsilon(\cdot)$ differs from $\hat{u}_o(\cdot)$ on a set of measure $O(\beta + \varepsilon)$.Let ε and β

be chosen such that $\varepsilon = 2c_1(\theta + 1)\beta/\alpha$. Then (69) holds and

$$\text{meas}\left\{ t \in [0,1] \,,\ u^{\varepsilon}_{\beta}(t) \neq \hat{u}_o(t) \right\} = O(\beta) \ .$$

Using Lemma 3.2 we obtain that the trajectory $(x_\beta(\cdot), y_\beta(\cdot))$, corresponding to $u^{\varepsilon}_{\beta}(\cdot)$ satisfies

$$\| x_\beta - \hat{x}_o \|_C = O(\beta) \quad \text{and} \quad \| y_\beta - \hat{y}_o \|_{L^1} = O(\beta + \omega(\hat{u}_o, \beta)_1) \ .$$

Since

$$\hat{I}_\beta \leqslant I(x_\beta(\cdot), y_\beta(\cdot), u^{\varepsilon}_{\beta}(\cdot)) \ ,$$

the upper estimate follows from the Lipschitz continuity of $f(\cdot)$.

Now, let $x^{\beta}_{o}(\cdot)$ be the solution of (59) for $\hat{u}_\beta(\cdot)$. By Lemma 3.2

$$|x^{\beta}_{o}(1) - x^1| = O(\beta) \ .$$

Applying Lemma 3.3 we select a feasible control $\bar{u}_\beta(\cdot)$ driving the state of (62a) to x^1 at $t=1$ such that

$$\text{meas}\left\{ t \in [0,1] \,,\ \bar{u}_\beta(t) \neq \hat{u}_\beta(t) \right\} = O(\beta)$$

and

$$\omega(\bar{u}_\beta, \beta)_1 \leqslant \omega(\hat{u}_\beta, \beta)_1 + c_5 \beta \ .$$

If $(\bar{x}_\beta(\cdot), \bar{y}_\beta(\cdot))$ solves (59) for $\bar{u}_\beta(\cdot)$, then by Lemma 3.2 we have

$$\| \bar{x}_\beta - \hat{x}_\beta \|_C = O(\beta) \ ,$$

$$\| \bar{y}_\beta - \hat{y}_\beta \|_{L^1} = O(\beta + \omega(\bar{u}_\beta, \beta)_1) = O(\beta + \omega(\hat{u}_\beta, \beta)_1) .$$

From

$$\hat{I}_o \leqslant I(\bar{x}_\beta(\cdot), \bar{y}_\beta(\cdot), \bar{u}_\beta(\cdot))$$

we get the lower estimate, Q.E.D.

<u>Corollary 3.1</u>. Let the function $f(\cdot)$ be differentiable with respect to (x,y,u) and its derivatives be continuous. Let

$$f_o(u,t) = f(\hat{x}_o(t), -A_4^{-1}(t)(A_3(t)\hat{x}_o(t) + B_2(t)u), u, t)$$

be strongly convex with respect to u uniformly in $t \in [0,1]$, and the variations of the matrix $B_o(t)$ and of the function $\frac{\partial}{\partial u} f_o(u,t)$ on $[0,1]$ be bounded uniformly in $u \in V$. Then $\hat{I}_\beta - \hat{I}_o \leqslant c_6 \beta$.

Proof. Using the strong concavity of the Hamiltonian and the Maximum principle as in § 2.3.4 one can show that the variation of $\hat{u}_o(\cdot)$ is bounded. Then the desired estimate follows from Theorem 3.7.

Corollary 3.2. Let

$$f(x,y,u,t) = f_1(x,u,t) + f_2(x,t)y ,$$

where $f_1(\cdot)$ has continuous derivatives with respect to x and u and is strongly convex with respect to u uniformly in x,t. The function $f_2(\cdot)$ is continuously differentiable, the elements of $A_1(t)$ are in C^1, and the elements of $B_1(t)$ have bounded variations. Then

$$\hat{I}_\beta - \hat{I}_o \geqslant c_7 \beta \ln \beta .$$

Moreover, if $B_2(t) = 0$, then $I_\beta - I_o \geqslant -c_8 \beta$.

Proof. Since for small β the point $(x^1, y^1) \in \operatorname{int} K_\beta$, then the vectors determining the final values of the adjoint variables in the Maximum principle are bounded when $\beta \to 0$. The estimation of the adjoint variables leads to the equation

$$\beta \dot{z} = A_4(t) z + p_\beta(t) \quad , \; z(1) = q_\beta / \beta ,$$

where

$$\limsup_{\beta \to 0} \left(|q_\beta| + \max_{0 \leqslant t \leqslant 1+\beta \ln \beta} |\dot{p}_\beta(t)| \right) < +\infty .$$

The solution of this equation has bounded variation on the interval $[0,1+\beta\ln\beta]$. Then, using the strong concavity of the Hamiltonian as in § 2.3.4 one can get

$$\omega(\hat{u}_\beta, \beta)_1 \leqslant \omega(\hat{u}_\beta, \beta)_{L^1(0,1+\beta\ln\beta)} - 2\beta\ln\beta \|\hat{u}_\beta\|_L \leqslant -c_9 \beta \ln \beta .$$

If $B_2(t) = 0$, the "fast" adjoint variables are not involved directly in the Maximum principle. In this case the estimate $O(\beta)$ is exact in the class of problems considered.

Remark 3.3. One can easily adapt (and simplify) the presented technique in order to estimate the performance convergence of regularly perturbed optimal control problems with fixed final state and constrained controls.

3.5.2. Estimates of the optimal control

Here we apply directly the analysis of Chapter 1 to the following singularly perturbed problem

$$J_\beta(u(\cdot)) = g(x(1)) + \int_0^1 (f(x(t),y(t),t) + h(u(t),t))dt \rightarrow \inf \quad (70)$$

subject to

$$\dot{x} = A_1(t)x + A_2(t)y + B_1(t)u \ , \ x(0) = x^o \ ,$$
$$\beta\dot{y} = A_3(t)x + A_4(t)y + B_2(t)u \ , \ y(0) = y^o \ , \quad (71)$$

$$u(\cdot) \in U = \{ u(\cdot) \in L^2(R^r), \ u(t) \in V \text{ for a.e. } t \in [0,1] \} . \quad (72)$$

We assume that

A10. The matrices $A_i(\cdot)$ are C^1 and $B_j(\cdot)$ are Lipschitz continuous. The eigenvalues of the matrix $A_4(t)$ have negative real parts for all $t \in [0,1]$. The set V is closed and convex. The function $g(\cdot)$ is convex and C^1, the derivative $\frac{d}{dx}g(\cdot)$ is Lipschitz continuous with a constant L_x. The function $f(\cdot)$ is continuous, $f(\cdot,t)$ is convex and differentiable with continuous derivatives. The derivatives $\frac{\partial}{\partial x}f(\cdot)$ and $\frac{\partial}{\partial u}f(\cdot)$ are Lipschitz continuous with respect to x and y with Lipschitz constants L_x and L_y respectively, uniformly in $t \in [0,1]$. The function $h(\cdot)$ is continuous, it is strongly convex with respect to u uniformly in t and the derivative $\frac{\partial}{\partial u}h(\cdot)$ is Lipschitz continuous.

The reduced problem, corresponding to $\beta = 0$, consists in minimizing (70) subject to (72) and

$$\dot{x} = A_o(t)x + B_o(t)u \ , \ x(0) = x^o \ ,$$

$$y(t) = - A_4^{-1}(t)(A_3(t)x(t) + B_2(t)u(t)), \ t \in [0,1] \ , \quad (73)$$

where $A_o = A_1 - A_2A_4^{-1}A_3$ and $B_o = B_1 - A_2A_4^{-1}B_2$.

The optimal control $\hat{u}_\beta(\cdot)$ exists and is unique for all $\beta \geq 0$. By repeating the arguments of Theorem 2.8 one can define $\hat{u}_\beta(\cdot)$ as a Lipschitz continuous function of the time $t \in [0,1]$ for every $\beta \geq 0$.

Lemma 2.10 yields that the functional $J_\beta(\cdot)$ is Frechet differentiable in $L^2(R^r)$ and the derivative $J'_\beta(\cdot)$ is given by

$$(J'_\beta(u(\cdot)))(t) = -\frac{\partial}{\partial u} H_\beta(u(t),t) ,$$

where

$$H_\beta(u,t) = - h(u,t) + (p_\beta^T(t)B_1(t) + q_\beta^T(t)B_2(t))u ,$$

and $(p_\beta(\cdot),q_\beta(\cdot))$ solves the adjoint equation

$$\dot p = - A_1^T(t)p - A_3^T(t)q + \frac{\partial}{\partial x} f(x(t),y(t),t) , \quad p(1) = -\frac{d}{dx} g(x(1)),$$

$$\beta\dot q = - A_2^T(t)p - A_4^T(t)q + \frac{\partial}{\partial y} f(x(t),y(t),t) , \quad q(1) = 0 . \tag{74}$$

Similarly, for the limit problem we have

$$(J'_o(u(\cdot)))(t) = -\frac{\partial}{\partial u} H_o(u(t),t) ,$$

where

$$H_o(u,t) = - h(u,t) + (p_o^T(t)B_1(t) + q_o^T(t)B_2(t))u ,$$

and $p_o(\cdot)$ solves the first equation in (74) while $q_o(\cdot)$ is defined by the second one with $\beta = 0$, that is

$$q_o(t) = - A_4^{-1}(t)^T(A_2(t)^T p_o(t) - \frac{\partial}{\partial y} f(x(t),y(t),t)) .$$

We remind that, denoting by $Y(t,s,\beta)$ the fundamental matrix solution of $\beta\dot z = A_4(t)z$, there exist constants $\sigma_o, \sigma > 0$ such that

$$|Y(t,s,\beta)| \leqslant \sigma_o \exp(-\sigma \frac{t-s}{\beta})$$

for all $t,s \in [0,1]$, $t \geqslant s$.

Denote

$$R_1 = \sigma_o(\| \frac{d}{dt}(A_2A_4^{-1}) \| \|A_3\| + \|A_2A_4^{-1}\|_C \|A_3\|_C)/\sigma ,$$

$$R_3 = \sigma_o(\| \frac{d}{dt}(A_4^{-1}A_3) \| \| A_2\| + \|A_4^{-1}A_3\|_C \|A_2\|_C)/\sigma .$$

Let $\beta_1 \in (0, \min\{1/R_1, 1/R_3\})$.

Theorem 3.8. For every $\beta \in (0,\beta_1)$

$$\|\hat{u}_\beta - \hat{u}_o\| \leqslant c_1\sqrt{\beta} + c_2\beta , \tag{75}$$

where the constants c_1 and c_2 are evaluated in the proof.

Proof.We apply Corollary 1.3.Let $\Delta x = x_\beta - \hat{x}_o$, $\Delta y = y_\beta - \hat{y}_o$, where $(x_\beta(\cdot),y_\beta(\cdot))$ is determined by (71) for $\hat{u}_o(\cdot)$.Then

$$\Delta x(t) = \int_0^t (A_1(s)\Delta x(s) + A_2(s)\Delta y(s))\, ds , \tag{76}$$

$$\Delta y(t) = Y(t,0,\beta)y^o + \frac{1}{\beta}\int_0^t (Y(t,s,\beta)(A_3(s)\Delta x(s) - A_4(s)\hat{y}_o(s))ds - \hat{y}_o(t) . \tag{77}$$

Since $\hat{y}_o(\cdot)$ is Lipschitz continuous, integrating by parts we have

$$\left|\frac{1}{\beta}\int_0^t Y(t,s,\beta)A_4(s)\hat{y}_o(s)ds + \hat{y}_o(t)\right|$$

$$\leqslant \left|\int_0^t \frac{\partial}{\partial s}Y(t,s,\beta)\hat{y}_o(s)ds - y_o(t)\right| \leqslant \sigma_o\exp(-t/\beta)|\hat{y}_o(0)| + \frac{L_o\sigma_o}{\sigma}\beta,$$

where L_o is the Lipschitz constant of $\hat{y}_o(\cdot)$.Furthermore

$$\frac{1}{\beta}\int_0^t A_2(s)\int_0^s Y(s,v,\beta)A_3(v)\Delta x(v)dvds$$

$$= \int_0^t A_2(s)A_4^{-1}(s)\frac{\partial}{\partial s}\int_0^s Y(s,v,\beta)A_3(v)\Delta x(v)dvds$$

$$- \int_0^t A_2(s)A_4^{-1}(s)A_3(s)\Delta x(s)ds = A_2(t)A_4^{-1}(t)\int_0^t Y(t,s,\beta)A_3(s)\Delta x(s)ds$$

$$- \int_0^t A_2(s)A_4^{-1}(s)A_3(s)\Delta x(s)ds$$

$$+ \int_0^t \frac{d}{ds}(A_2(s)A_4^{-1}(s))\int_0^s Y(s,v,\beta)A_3(v)\Delta x(v)dsdv .$$

Using the last two relations in (76) we get

$$|\Delta x(t)| \leq \|A_o\|_C \int_0^t |\Delta x(s)|\,ds + \beta(R_1 \|\Delta x\|_C + \frac{\sigma_o}{\sigma}\|A_2\|_C(|\hat{y}_o(0)| + L_o + |y^o|)).$$

By the Gronwall lemma we conclude that

$$\|\Delta x\|_C \leq (R_2 e^a/(1 - R_1\beta_1))\beta = d_1\beta , \tag{78}$$

where $a = \|A_o\|_C$, $R_2 = \sigma_o\|A_2\|_C(|\hat{y}_o(0)| + L_o + |y^o|)/\sigma$. Using this estimate in (77) we obtain

$$\|\Delta y\| \leq \sigma_o(|y^o| + |\hat{y}_o(0)|)\sqrt{\beta}/\sqrt{2\sigma} + (L_o\sigma_o/\sigma + \sigma_o\|A_3\|_C d_1/\sigma)\beta$$

$$= d_2\sqrt{\beta} + d_3\beta, \tag{79}$$

where d_2 and d_3 are determined in appropriate way.

Let L_1 be the Lipschitz constant of $q_o(\cdot)$(we denote by $(p_o(\cdot), q_o(\cdot))$ the optimal adjoint trajectory for the reduced problem). Denote

$$R_4 = 2L_x d_1 + \sigma_o\|A_3\|_C(|q_o(1)| + L_1)/\sigma + \sigma_o\|A_3\| L_y d_3/\sigma ,$$

$$R_5 = \sigma_o\|A_3\| L_y d_2/\sigma ,$$

$$d_4 = R_5 e^a/(1 - R_3\beta_1) , \quad d_5 = R_4 e^a/(1 - R_3\beta_1) ,$$

$$d_6 = \sigma_o|q_o(1)|/\sqrt{2\sigma} + \sigma_o(\|A_3\| d_4 + L_y d_2)/\sigma ,$$

$$d_7 = \sigma_o(L_1 + \|A_3\| d_5 + L_y d_3)/\sigma .$$

Let $(\tilde{p}_\beta(\cdot),\tilde{q}_\beta(\cdot))$ be the solution of (74) for $(x_\beta(\cdot),y_\beta(\cdot))$ and let $\Delta p = \tilde{p}_\beta - p_o$, $\Delta q = \tilde{q}_\beta - q_o$. Proceeding in the same manner as for the state we get

$$\|\Delta p\| \leq d_4\sqrt{\beta} + d_5\beta , \quad \|\Delta q\| \leq d_6\sqrt{\beta} + d_7\beta .$$

Then, by Corollary 1.3 from Chapter 1

$$\|\hat{u}_\beta - \hat{u}_o\| \leq (\|B_1\|\|\Delta p\| + \|B_2\|\|\Delta q\|)/2\varkappa ,$$

where $\varkappa$ is the constant of the strong convexity of $h(\cdot,t)$. Thus, denoting

3.6. Final remarks

Doubtless, the most important and, probably, the most difficult problem arising from the analysis in this chapter, is the problem of generalizing the presented results to systems nonlinear with respect to the fast states. In this connection it would be relevant to find an analogue of Tichonov's theorem for singularly perturbed differential inclusions

$$\begin{bmatrix} \dot{x} \\ \beta\dot{y} \end{bmatrix} \in F(x,y,t) \quad .$$

As already mentioned in Remark 3.1, the set of solutions of this inclusion does not converge pointwise to the set of solutions of the reduced inclusion

$$\begin{bmatrix} \dot{x} \\ 0 \end{bmatrix} \in F(x,y,t) \quad .$$

Probably, the concept of invariance in differential inclusions, see Peng [59] , would play an important role in such a generalization.

The next open problem is related to the state constraints. Example 3.2 showed that the fast state constraint "removes" the singularity to the slow system and changes essentially the reachable set of the full-order system. For some special sequences β_k, for example such that

$$\lim_{k\to\infty} (\beta_{k+1} - \beta_k)/\beta_k = 0 \ ,$$

one can prove that the sequence of the reachable sets is a fundamental sequence in the Hausdorff metric, hence it has a limit set.

Further generalizations of the presented results are concerned with the spectrum of the matrix $A_4(t)$. Suppose that a part of the eigenvalues of this matrix have strictly negative real parts and the remaining have strictly positive real parts for all $t \in [0,1]$. Consider the Lagrange problem from § 3.4.2. If the matrices are time-constant one can decompose the system into two subsystems associated with the positive and the negative real parts of the eigenvalues, respectively. Then the first subsystem can be regarded as a system with fixed final state and we shall try to reach the given initial state, changing the direction of the time. To the second system, associated with the eigenvalues with negative real parts we apply the usual scheme. Some results in this direction can be found in Gičev [27].

A typical difficulty arising in the convergence analysis of optimal state and control for singularly perturbed problems is to choose a suitable matric,describing the boundary layer effects.Clearly, the L^2 metric used in Theorem 3.8 is too weak.On the other hand, the estimations in the uniform metric require rather restrictive conditions,as in Theorem 3.9.A more natural metric, associated with the behaviour of the fast states is the Hausdorff distance between functions.In this direction, the results in Sendov [68] could be very helpfull.

In this chapter we confined ourselves to singular perturbation problems ,for which the singularity is represented by a small parameter in the derivative of the state.The reader can find results, concerned with time delay perturbations of hereditary control systems in Gičev and Dontchev [29] and Dontchev [15] .

CHAPTER 4

FINITE - DIFFERENCE APPROXIMATIONS

4.1. Introduction

The numerical procedures for solving optimal control problems often use finite-difference approximations of the functions and operators forming the problem.In this way one obtains a finite-dimensional mathematical programming problem depending on the parameter of the discretization.Actually, this is a change of the model, therefore, the evaluation of the discretization error can be generally regarded as a sensitivity problem, for which the perturbation parameter is represented by the discretization step.The finite-difference approximation will be well-posed when the solution (optimal value, optimal control) of the approximate problem converges to the solution of the original one.Furthermore, it is not less important than the convergence itself to estimate the convergence rate as a function of the discretization parameter.

The performance convergence of discrete approximations to optimal control problems has been studied recently by Budak and Vasilev [7], Culum [10], Ermolev et al.[23], Levikov [47], Malanowski [51], and Mordukhovič [56].Hager [37] and Reddien [63] estimated the optimal control error for high order integration schemes applied to unconstrained control pronlems.Malanowski [53] examined the approximation error for abstract strictly convex problems with constraints.In a different direction, dual approximations to constrained problems were developed in Hager [35],[36] and Dontchev [13].

This chapter deals with the convergence rate of the finite-difference approximation provided by Euler integraion scheme to constrained optimal control problems.The error is represented by the distance between the discrete optimal control, considered as a step function, and the optimal control of the continuous problem As it is known from the numerical analysis, the closeness of the approximating discrete control to the continuous one is limited by the order of smoothness of the optimal control as a function of the time.Therefore, the crucial role in the error evaluation is played by the regularity of the optimal control.The continuity properties of the optimal control have been studied recently by Hager [38] and Malanowski [51] .

Section 4.2 considers a problem with local inequality constraints, similar to that of Section 2.3.After a series of auxliary lemmas,the main result is established: first order convergence of the optimal controls in the L^2 metric.This result was obtained independently by Malanowski [52] and Dontchev [14] , by different proof.We present here the proof from the author's paper [14] .In § 4.2.2 we refine this estimate to the unitorm metric assuming that there are no state constraints.Paragraph 4.2.3 presents estimations for the dual variables. Our presentation follows the scheme ot Section 2.3, although the proofs use different ideas.

Section 4.3 deals with the optimal control convergence for a problem with mixed integral constraints.As an example we consider a fixed final state problem.

4.2. Problems with local inequality constraints

4.2.1. State and control constraints

The following optimal control problem is considered :

$$I(x(\cdot),u(\cdot)) = \int_0^1 f(x(t),u(t),t)dt \rightarrow \inf \tag{1}$$

subject to the constraints

$$\dot{x} = A(t)x + B(t)u \quad ,x(0) = x^0 , \tag{2}$$

$$\Theta(x(t),t) \leqslant 0 \tag{3}$$

$$\Psi(u(t),t) \leqslant 0 \quad , \text{ for } t \in [0,1] , \tag{4}$$

$$u(\cdot) \in L^2(R^m) \;, \; x(\cdot) \in A(R^n) \;,$$

where $\Theta(\cdot)$: $R^n \times [0,1] \rightarrow R^p$, $\Psi(\cdot)$: $R^m \times [0,1] \rightarrow R^q$.

We assume that:

A1. The elements of the matrices $A(t)$ and $B(t)$ are Lipschitz continuous on $[0,1]$. The functions $f(\cdot),\theta(\cdot),\Psi(\cdot)$ and $\frac{\partial}{\partial x}\theta(\cdot)$ are C^2.

A2. The functions $f(\cdot,t)$ and the components of $\theta(\cdot,t),\Psi(\cdot,t)$ are convex for all $t\in[0,1]$. There exists $\alpha>0$ such that

$$z^T\frac{\partial^2}{\partial z^2}f(y,t)z \geqslant \alpha|u|^2$$

for all $y,z\in R^{m+n}$ and $t\in[0,1]$.

A3. There exist a continuous control $\bar{u}(\cdot)$ and a constant $\beta<0$ such that

$$\theta(\bar{x}(t),t)_i \leqslant \beta \quad , \quad \Psi(\bar{u}(t),t)_j \leqslant \beta$$

for all $t\in[0,1]$, $i = 1,\ldots,p$, $j = 1,\ldots,q$, where $\bar{x}(\cdot)$ corresponds to $\bar{u}(\cdot)$.

Denote by $(\hat{x}(\cdot),\hat{u}(\cdot))$ the optimal solution, which, clearly, exists and is unique. We assume also that Hager's regularity condition from § 2.3.3 holds, that is

A4. There exists $\gamma>0$ such that for all $t\in[0,1]$ and for all z

$$\left|\left[G_c^T(t),B^T(t)G_s^T(t)\right]z\right| \geqslant \gamma|z| \quad ,$$

where $G_c(t)$ is the matrix, which rows are the derivatives of the components of $\psi(\cdot,t)$ with respect to u at $\hat{u}(t)$ corresponding to binding constraints for $\hat{u}(t)$. The matrix $G_s(t)$ is defined similarly.

The results from Hager [38] and Hager and Mitter [39], involved in our analysis are joined in the following theorem, see also Theorem 2.3:

<u>Theorem 4.1</u>. There exist an optimal control $\hat{u}(\cdot)$, a corresponding trajectory $\hat{x}(\cdot)$, and optimal dual multipliers $q(\cdot),\nu(\cdot),\lambda(\cdot)$, $q(1)=0$ $\lambda(\cdot)\geqslant 0$, $\nu(\cdot)$ is nondecreasing on $[0,1]$, $\nu(1)=0$, such that $\hat{x}(\cdot)$, $\dot{q}(\cdot),\hat{u}(\cdot),\nu(\cdot),\lambda(\cdot)$ are Lipschitz continuous on $[0,1)$. The following optimality conditions are satisfied

$$\frac{\partial}{\partial u}f(\hat{x}(t),\hat{u}(t),t) + B^T(t)(q(t) - \frac{\partial}{\partial x}\theta^T(\hat{x}(t),t)\nu(t))$$

$$+ \frac{\partial}{\partial u}\psi^T(\hat{u}(t),t)\lambda(t) = 0 \quad , \tag{5}$$

$$\dot{q} = -A^T(t)(q - \frac{\partial}{\partial x}\theta^T(\hat{x}(t),t)\nu(t)) - \frac{\partial}{\partial u}f(\hat{x}(t),\hat{u}(t),t)$$

$$+ \frac{d}{dt}(\frac{\partial}{\partial x}\theta^T(\hat{x}(t)t))\nu(t) , \qquad (6)$$

$$[\nu,\theta(\hat{x})] = \langle \lambda,\Psi(\hat{u})\rangle = 0 . \qquad (7)$$

Let us introduce the uniform grid $\{ih\}$, $i = 0,\dots,N$, $h = 1/N$. The following discrete approximation to the problem (1) - (4) is examined:

$$I_N(x,u) = \sum_{i=0}^{N-1} hf(x_i,u_i,t_i) \longrightarrow \inf , \qquad (8)$$

$$x_{i+1} = (I + hA_i)x_i + hB_iu_i , \quad i = 0,\dots,N-1 , \qquad (9)$$

$$x_0 = x^0 ,$$

$$\theta(x_i,t_i) \leqslant 0 , \quad i = 1,\dots,N , \qquad (10)$$

$$\Psi(u_i,t_i) \leqslant 0 , \quad i = 0,\dots,N-1 , \qquad (11)$$

where $A_i = A(t_i)$, $B_i = B(t_i)$, $i = 0,\dots,N$, I is the identity.

Let us recall that the normal Lagrange functional for the discrete problem (8) - (11) has the form

$$L_N(x,u;p,\mu,\lambda) = I_N(x,u) + \sum_{i=0}^{N-1} (p_i^T(x_{i+1} - (I + hA_i)x_i + hB_iu_i)$$

$$+ \theta^T(x_{i+1},t_{i+1})\mu_{i+1} + \Psi^T(u_i,t_i)\lambda_i) , \qquad (12)$$

where $p_i \in R^n$, $\mu_i \in R^p$, $\mu_i > 0$, $\lambda_i \in R^q$, $\lambda_i \geqslant 0$, $i = 0,\dots,N$.

By the classical convex programming we get:

Theorem 4.2. Suppose that h is sufficiently small.Then there exist unique solution $(\hat{x}^N,\hat{u}^N)$ to (8) - (11) and dual multipliers p^N, μ^N,λ^N such that

$$\hat{I}_N = I_N(\hat{x}^N,\hat{u}^N) = \min L_N(x,u;p^N,\mu^N,\lambda^N) , \quad x_0 = x^0, \ x \in R^{Nn}, u \in R^{Nm},$$

and the following relations hold at $(\hat{x}^N, \hat{u}^N, p^N, \mu^N, \lambda^N)$:

$$h \frac{\partial}{\partial u} f(x_i, u_i, t_i) - hB_i^T p_i + \frac{\partial}{\partial u} \Psi^T(u_i, t_i)\lambda_i = 0 , \qquad (13)$$

$$p_{i-1} = (I + hA_i^T)p_i - h \frac{\partial}{\partial x} f(x_i, u_i, t_i) - \frac{\partial}{\partial x} \Theta^T(x_i, t_i)\mu_i ,$$

$$p_{N-1} + \frac{\partial}{\partial x} \Theta^T(x_N, t_N)\mu_N = 0 , \qquad (14)$$

$$\sum_{i=1}^{N} \Theta^T(x_i, t_i)\mu_i = \sum_{i=0}^{N-1} \Psi^T(u_i, t_i)\lambda_i = 0 . \qquad (15)$$

In the sequel it is assumed that h is sufficiently small.The optimal discrete solution $(\hat{x}^N, \hat{u}^N)$ is considered as a step function across the grig points t_i.

The main result of this paragraph follows:

Theorem 4.3. The following estimation holds

$$\|\hat{x}^N - \hat{x}\|_C + \|\hat{u}^N - \hat{u}\| = O(h) .$$

The proof will be completed as a series of lemmas.

Lemma 4.1. The sequence $\hat{I}_N$ and $\hat{u}^N$ are bounded when $N \to +\infty$ and there exists a constant c such that

$$\|\hat{x}^N - \hat{x}\|_C \leqslant c(\|\hat{u}^N - \hat{u}\| + h). \qquad (16)$$

Proof. The first statement follows from the strong convexity of the functional (see Lemma 2.2).The estimate (16) is obtained by the Gronwall lemma, using the Lipschitz continuity of $\hat{u}(\cdot)$.

Lemma 4.2.

$$\limsup_{N \to \infty} \Big(\max_{0 \leqslant i \leqslant N-1} (|\hat{u}_i^N| + |p_i^N|) + \sum_{i=0}^{N-1} (\mu_{i+1}^N + \lambda_i^N) \Big) < +\infty .$$

Proof.The proof goes parallelly to the proof of Lemma 2.3.Let $\bar{x}_i$ solves the discrete equation (9) for $\bar{u}_i = \bar{u}(t_i)$, i=0,...,N. Clearly, $\bar{x}_i \to \bar{x}(t_i)$ as $N \to +\infty$ uniformly in i.Then, for sufficiently large N we have

$$\Theta(\bar{x}_i, t_i)_k \leqslant \beta/2 , \quad i= 1,\dots,N, \; k=1,\dots,p .$$

By the duality relation in Theorem 4.2 we conclude that

$$\hat{I}_N - I_N(\bar{x},\bar{u}) \leqslant \frac{\beta}{2}\sum_{i=0}^{N-1}\Big(\sum_{j=1}^{p}\mu^N_{i+1,j} + \sum_{j=1}^{q}\lambda^N_{ij}\Big) \leqslant 0$$

Taking into account Lemma 4.1,this implies

$$\limsup_{N\to\infty}\sum_{i=0}^{N-1}\big(|\mu^N_{i+1}| + |\lambda^N_i|\big) < +\infty .$$

Then, applying the discrete Gronwall lemma to (14) we get

$$\limsup_{N\to\infty}\max_{0\leqslant i\leqslant N-1}|p^N_i| < +\infty .$$

If

$$H_N(u,t_i) = -f(\hat{x}^N_i,u,t_i) + (p^N_i)^T B_i u ,$$

then

$$\alpha|\hat{u}^N_i - \bar{u}_i| \leqslant \Big|\frac{\partial}{\partial u}H_N(\bar{u}_i,t_i)\Big| , \quad i = 0,\dots,N-1 ,$$

hence $\limsup_{N\to\infty}\max_{0\leqslant i\leqslant N-1}|\hat{u}^N_i| < +\infty$,Q.E.D.

Lemma 4.3.

$$\|\hat{x}^N - \hat{x}\|_C + \|\hat{u}^N - \hat{u}\| = O(\sqrt{h}) .$$

Proof. Let $x^N(t) = \hat{x}(t_i)$, $u^N(t) = \hat{u}(t_i)$, $t\in[t_i,t_{i+1})$, $i = 0,\dots,N-1$, $x^N_N = \hat{x}(1)$.We have

$$I_N \leqslant L_N(x^N,u^N;p^N,\mu^N,\lambda^N)$$

$$\leqslant I_N(x^N,u^N) + \sum_{i=0}^{N-1}(p^N_i)^T(x^N_{i+1} - (I + hA_i) - hB_i u^N_i). \qquad (17)$$

Since $\hat{u}(\cdot)$ is Lipschitz continuous

$$\max_{0\leqslant i\leqslant N-1}|x^N_{i+1} - (I + hA_i)x^N_i - hB_i u^N_i| = O(h^2) . \qquad (18)$$

The function

$$p(t) = \frac{\partial}{\partial x}\theta^T(x(t),t)\nu(t) - q(t)$$

is Lipschitz continuous on $[0,1)$. Let $p_i = p(t_i)$, $i = 0,\dots,N$. Then, from (6) and Theorem 4.1 it follows that

$$\sum_{i=1}^{N-1} \Big|p_{i-1} - (I + hA_i^T)p_i + h\frac{\partial}{\partial x}f(x_i^N,u_i^N,t_i)$$

$$+ \frac{\partial}{\partial x}\theta^T(x_i^N,t_i)(\nu(t_i) - \nu(t_{i-1}))\Big| = O(h) \;, \qquad (19)$$

Using the strong convexity we have

$$\hat{I}_N = L_N(\hat{x}^N,\hat{u}^N;p^N,\mu^N,\lambda^N) \geqslant I_N + \sum_{i=0}^{N-1} p_i^T(\hat{x}_{i+1}^N - (I + hA_i)\hat{x}_i^N$$

$$- hB_i\hat{u}_i^N) + \sum_{i=0}^{N} \theta^T(x_i^N,t_i)(\nu(t_i) - \nu(t_{i-1}))$$

$$\geqslant I_N(x^N,u^N) + \sum_{i=0}^{N-1} p_i^T(x_{i+1}^N - (I + hA_i)x_i^N - hB_iu_i^N)$$

$$+ \sum_{i=1}^{N} \theta^T(x_i^N,t_i)(\nu(t_i) - \nu(t_{i-1})) + \sum_{i=0}^{N-1} h\Big(\frac{\partial}{\partial u}f(x_i^N,u_i^N,t_i)$$

$$- B_i^Tp_i\Big)^T(\hat{u}_i^N - u_i^N) + \alpha\sum_{i=0}^{N-1} h\,|\hat{u}_i^N - u_i^N|^2$$

$$+ \sum_{i=0}^{N-1}\Big(p_{i-1} - (I + hA_i^T)p_i + h\frac{\partial}{\partial x}f(x_i^N,u_i^N,t_i)$$

$$+ \frac{\partial}{\partial x}\theta^T(x_i^N,t_i)(\nu(t_i) - \nu(t_{i-1}))\Big)^T(\hat{x}_i^N - x_i^N)$$

$$+ q(t_{N-1})^T(\hat{x}_{N-1}^N - x_{N-1}^N) \;. \qquad (20)$$

Let $\nu^N(t) = \nu(t_i)$ for $t\in[t_i,t_{i+1})$, $i = 0,\dots,N-1$, $\nu^N(1) = 0$. Applying Lemma 5.2 from Hager [36] we obtain

$$\sum_{i=1}^{N} \theta^T(x^N{}_i,t_i)(\nu(t_i)-\nu(t_{i-1})) = \int_0^1 \theta^T(\hat{x}(t),t)d(\nu^N(t)-\nu(t)) = O(h^2). \qquad (21)$$

The optimality condition (5) yields

$$\left(\frac{\partial}{\partial u} f(x_i^N, u_i^N, t_i) - B_i^T p_i\right)^T (\hat{u}_i^N - u_i^N) \geq 0, i = 0,\dots,N-1. \quad (22)$$

From the condition $q(1) = 0$ and Lemma 4.1 we get

$$q(t_{N-1})^T(\hat{x}_{N-1}^N - x_{N-1}^N) \geq -ch(\|\hat{u}^N - \hat{u}\| + h). \quad (23)$$

Combining (17) and (20) and using (18).(19),(21),(22) and (23) we obtain finally

$$\|\hat{u}^N - \hat{u}\|^2 \leq ch\left(\sum_{i=0}^{N-1} h|p_i^N - p(t_i)| + \|\hat{u}^N - \hat{u}\| + h\right). \quad (24)$$

Applying lemmas 4.1 and 4.2 we complete the proof.

As in § 2.3.3 we introduce the sets of δ-binding constraints

$$r_\delta(t) = \{ j \in \{1,\dots,p\} , \Theta(\hat{x}(t),t)_j \geq -\delta \} ,$$

$$c_\delta(t) = \{ j \in \{1,\dots,q\} , \Psi(u(t),t)_j \geq -\delta \} .$$

From Lemma 2.5 it follows that there exist $\delta_o > 0$ and $\gamma_o > 0$ such that for every $z = (z^1,z^2)$ and $t \in [0,1]$

$$\left| \sum_{j\in c(t)} \frac{\partial}{\partial u}\Psi(\hat{u}(t),t)_j z_j^1 + \sum_{j\in r(t)} (B^T(t) \frac{\partial}{\partial x}\Theta^T(\hat{x}(t),t))_j z_j^2 \right|$$

$$\geq \gamma_o(|z^1| + |z^2|) , \quad (25)$$

where $c(t)$ and $r(t)$ correspond to δ_o.

Let $n(t) = \{1,\dots,p\} \setminus r(t)$,$m(t) = \{1,\dots,q\} \setminus c(t)$ and $r_i = r(t_i)$, $c_i = c(t_i)$,$n_i = n(t_i)$,$m_i = m(t_i)$.In the sequel vectors (matrices) with superscripts $r(t)$,$c(t)$,$n(t)$,$m(t)$ denote subvectors (submatrices) associated with these sets of constraints.

<u>Lemma 4.4</u>. Let $\mu^N(t) = \mu_i^N$ for $t \in [t_{i-1},t_i)$, $i = 1,\dots,N$.There exists N_o such that for all $N > N_o$

$$\sup_{0\leq t\leq 1} (\mu^N)^{n(t)}(t) = 0.$$

Proof.Assume the opposite.Then there exist a sequence t^N ,

$t^N \to t' \in [0,1]$ as $N \to \infty$, and an index $j' \in n(t^N)$ such that $\mu^N_{j'}(t^N) > 0$ for all N.The complementary slackness condition (15) implies that $\theta^N(t^N) = 0$,where $\theta^N(t) = \Theta(\hat{x}^N_i, t_i)_{j'}$ for $t \in [t_{i-1}, t_i)$, $i = 1,\ldots,N$. On the other hand,by Lemma 4.3

$$0 = \lim_{N\to\infty} \theta^N(t^N) = \Theta(\hat{x}(t'),t')_{j'} .$$

But

$$\Theta(\hat{x}(t^N),t^N)_{j'} < -\delta_o ,$$

which contradicts the continuity of $\Theta(\cdot)$ and $\hat{x}(\cdot)$,Q.E.D.

Lemma 4.5.

$$\sum_{i=0}^{N-1} |(\lambda^N_i)^{m_i}| = O(h). \tag{26}$$

Proof.Using the notation of Lemma 4.3 we have

$$\hat{I}_N \leqslant I_N(x^N,u^N) + \sum_{i=0}^{N-1} (p^N_i)^T(x^N_{i+1} - (I + hA_i)x^N_i - hB_iu^N_i)$$
$$+ \sum_{i=0}^{N-1} \psi^T(u^N_i,t_i)\lambda^N_i .$$

Combining this inequality with (20) and using (18),(19),(21),(22) and (23) we obtain

$$\sum_{i=0}^{N-1} \psi^T(\hat{u}(t_i),t_i)\lambda^N_i \geqslant -ch .$$

Furthermore

$$-ch \leqslant \sum_{i=0}^{N-1} \psi^T(\hat{u}(t_i),t_i)\lambda^N_i \leqslant \sum_{i=0}^{N-1} \sum_{j\in m_i} \psi^T(\hat{u}(t_i),t_i)_j\lambda^N_{ij}$$
$$\leqslant -\delta_o \sum_{i=0}^{N-1} \sum_{j\in m_i} \lambda^N_{ij} \leqslant 0 .$$

This gives us (26),Q.E.D.

Lemma 4.6.

$$\sum_{i=0}^{N-1} |(\lambda_i^N)^{c_i}|^2 = O(h) . \qquad (27)$$

Proof. Let $C^N(t) = \frac{\partial}{\partial u}\psi(\hat{u}_i^N, t_i)^T$ for $t \in [t_i, t_{i+1}), i = 0, \ldots, N-1$ and $C(t) = \frac{\partial}{\partial u}\psi(\hat{u}(t), t)^T$ .Using Minkowski and Schwarz inequalities we get

$$\| C^c(\lambda^N)^c \| \leq \| C^c - (C^N)^c \| \| (\lambda^N)^c \| + \| (C^N)^c(\lambda^N)^c \| \quad .$$

Lemma 4.3 yields

$$\| C^c - (C^N)^c \| = O(\sqrt{h}) \ ,$$

and, by (25)

$$\gamma_0 \| (\lambda^N)^c \| \leq \| C^c(\lambda^N)^c \| \quad .$$

Hence

$$(\gamma_0 - c\sqrt{h}) \| (\lambda^N)^c \| \leq \| (C^N)^c(\lambda^N)^c \| \ . \qquad (28)$$

Furthermore, taking advantage of (13), lemmas 4.1, 4.2, 4.3 and 4.5 we conclude that

$$\| (C^N)^c(\lambda^N)^c \| \leq c(h + \| (\lambda^N)^m \|)$$

$$\leq c(h + \sqrt{h} \sum_{i=0}^{N-1} |(\lambda_i^N)^{m_i}|) = O(h) \ . \qquad (29)$$

Combining (28) and (29) we obtain (27), Q.E.D.

Lemma 4.7. Let $p(\cdot)$ be defined as in Lemma 4.3.Then

$$\sum_{i=0}^{N-1} h |p_i^N - p(t_i)| \leq c(\| \hat{u}^N - \hat{u} \| + h). \qquad (30)$$

Proof.Let us introduce the discrete variable ν^N as follows:

$$\nu_N^N = 0$$

$$\nu_{i-1,j}^N = \nu_{i,j}^N - \mu_{i,j}^N \quad \text{for } j \in r_i \ ,$$

$$\nu^N_{i-1,j} = \nu^N_{i,j} = \nu(t_i)_j \quad \text{for } j \in n_i \ , i = 1,\dots,N.$$

From Lemma 2.6 there exists $\gamma_1 > 0$ such that if $h < \gamma_1$ then ν^N_i is well-defined for all $i = 0,\dots,N$. Moreover, Lemma 4.4 implies that $\mu^N_i = \nu^N_i - \nu^N_{i-1}$ for all $i = 1,\dots,N$.

Let $S^N_i = \frac{\partial}{\partial x}\theta(\hat{x}^N_i,t_i)^T$, $C^N_i = C^N(t_i)$, $S(t) = \frac{\partial}{\partial x}\theta(x(t),t)^T$ and

$$q^N_i = S^N_i\nu^N_i - p^N_i \ , \ i = 0,\dots,N-1.$$

Then, by (13) and (14) we obtain the following relations

$$h\frac{\partial}{\partial u}f(\hat{x}^N_i,\hat{u}^N_i,t_i) + hB^T_i(q^N_i - S^N_i\nu^N_i) + C^N_i\lambda^N_i = 0 \ . \tag{31}$$

$$q^N_{i-1} = (I + hA^T_i)q^N_i + h\frac{\partial}{\partial x}f(\hat{x}^N_i,\hat{u}^N_i,t_i) - hA^T_iS^N_i\nu^N_i$$
$$+ (S^N_{i-1} - S^N_i)\nu^N_{i-1} \quad , \ q^N_{N-1} = (S^N_{N-1} - \dot{S}^N_N)\nu^N_{N-1} \ . \tag{32}$$

From (5) and (31) we get

$$|C(t_i)^{c_i}((\lambda^N_i)^{c_i} - h\,\lambda^{c_i}(t_i)) - hB^T_iS^{r_i}(t_i)((\nu^N_i)^{r_i} - \nu^{r_i}(t_i))|$$
$$\leqslant h|B_i||q^N_i - q(t_i)| + h|\frac{\partial}{\partial u}(f(\hat{x}^N_i,\hat{u}^N_i,t_i) - f(\hat{x}(t_i),\hat{u}(t_i),t_i))|$$
$$+ h|B_i|\,(|(S^N_i)^{r_i} - S^{r_i}(t_i)||(\nu^N_i)^{r_i}| + |(S^N_i)^{n_i} - S^{n_i}(t_i)||(\nu^N_i)^{n_i}|)$$
$$+ |(C^N_i)^{c_i} - C^{c_i}(t_i)||(\lambda^N_i)^{c_i}| + |(C^N_i)^{m_i}||(\lambda^N_i)^{m_i}| \ .$$

Let $\Delta q_i = q^N_i - q(t_i)$, $\Delta x_i = \hat{x}^N_i - \hat{x}(t_i)$, $\Delta u_i = \hat{u}^N_i - \hat{u}(t_i)$, $i = 0,\dots,N-1$. From Lemma 4.2 it follows that ν^N is bounded uniformly in $[0,1]$ when $N\to\infty$. The latter inequality and (25) yield

$$|(\lambda^N_i)^{c_i} - h\lambda^{c_i}(t_i)| + h|(\nu^N_i)^{r_i} - \nu^{r_i}(t_i)|$$
$$\leqslant c(h(|\Delta q_i| + |\Delta u_i| + |\Delta x_i|) + |\Delta u_i||(\lambda^N_i)^{c_i}| + |(\lambda^N_i)^{m_i}|) \ . \tag{33}$$

By the relation

$$\int_{t_{i-1}}^{t_i}\dot{S}(t)\nu(t)dt = (S(t_i) - S(t_{i-1}))\nu(t_i) + O(h^2) \ ,$$

and by (6) we obtain

$$q(t_{i-1}) = (I + hA_i^T)q(t_i) - hA_i^T S(t_i)\nu(t_i)$$
$$+ h\frac{\partial}{\partial x}f(\hat{x}(t_i),\hat{u}(t_i),t_i) + (S(t_{i-1}) - S(t_i))\nu(t_{i-1})$$
$$+ O(h^2)\ , \qquad q(t_{N-1}) = O(h)\ . \tag{34}$$

The following estimate will be used further on

$$|(S_i^N - S_{i-1}^N)\nu_{i-1}^N - (S(t_i) - S(t_{i-1}))\nu(t_{i-1})|$$
$$\leqslant |(S_i^N - S_{i-1}^N) - (S(t_i) - S(t_{i-1}))||\nu(t_{i-1})| + ch|\nu_i^N - \nu(t_{i-1})|$$
$$\leqslant h|\frac{\partial^2}{\partial x^2}\Theta(\hat{x}_{i-1}^N,t_{i-1})(A_{i-1}\hat{x}_{i-1}^N + B_{i-1}\hat{u}_{i-1}^N)$$
$$- \frac{\partial^2}{\partial x^2}\Theta(\hat{x}(t_{i-1}),t_{i-1})(A_{i-1}\hat{x}(t_{i-1}) + B_{i-1}\hat{u}(t_{i-1}))\|\nu(t_{i-1})|$$
$$+ h|\frac{\partial^2}{\partial x\partial t}(\Theta(\hat{x}_{i-1}^N,t_{i-1}) - \Theta(\hat{x}(t_{i-1}),t_{i-1}))\|\nu(t_{i-1})|$$
$$+ ch|\nu_{i-1}^N - \nu(t_{i-1})| + ch^2$$
$$\leqslant ch(|\Delta x_{i-1}| + |\Delta u_{i-1}| + |(\nu_{i-1}^N)^{r_{i-1}} - \nu^{r_{i-1}}(t_{i-1})| + h). \tag{35}$$

Combining (32) and (34) and using (35) we get

$$|\Delta q_{i-1}| \leqslant (1 + ch)|\Delta q_i| + ch(|\Delta x_i| + |\Delta u_i| + |\Delta x_{i-1}| + |\Delta u_{i-1}|$$
$$+ |(\nu_i^N)^{r_i} - \nu^{r_i}(t_i)| + |(\nu_i^N)^{r_{i-1}} - \nu^{r_{i-1}}(t_{i-1})| + h),$$
$$|\Delta q_{N-1}| = O(h)\ . \tag{36}$$

By (33) and the latter inequality we have

$$|\Delta q_{i-1}| \leqslant (1 + ch)|\Delta q_i| + c(h(|\Delta x_i| + |\Delta u_i| + |\Delta x_{i-1}| + |\Delta u_{i-1}|$$
$$+ h) + |\Delta u_i||(\lambda_i^N)^{c_i}| + |\Delta u_{i-1}||(\lambda_{i-1}^N)^{c_{i-1}}| + |(\lambda_i^N)^{m_i}|$$
$$+ |(\lambda_{i-1}^N)^{m_{i-1}}|\ .$$

Hence, by the discrete Gronwall lemma and Schwarz inequality

$$\max_{0\leqslant i\leqslant N-1} |\Delta q_i| \leqslant c(h + \sum_{i=0}^{N-1} h(|\Delta u_i| + |\Delta x_i|)$$

$$+ \sum_{i=0}^{N-1} (|\Delta u_i||(\lambda_i^N)^{c_i}| + |(\lambda_i^N)^{m_i}|))$$

$$\leqslant c(h + \|\hat{u}^N - \hat{u}\| + \|\hat{x}^N - \hat{x}\|_C$$

$$+ \|\hat{u}^N - \hat{u}\| (\sum_{i=0}^{N-1} |(\lambda_i^N)^{c_i}|^2)^{0.5}/h^{0.5} + \sum_{i=0}^{N-1} |(\lambda_i^N)^{m_i}|).$$

Thus, by lemmas 4.1, 4.5 and 4.6

$$\max_{0\leqslant i\leqslant N-1} |q_i| \leqslant c(\|\hat{u}^N - \hat{u}\| + h) . \tag{37}$$

Once again from lemmas 4.1, 4.5, 4.6 and from (33) we get

$$\sum_{i=0}^{N-1} h|\nu_i^N - \nu(t_i)| \leqslant c(\|\hat{u}^N - \hat{u}\| + h) . \tag{38}$$

From the definition of $p(\cdot)$, Lemma 4.1 and the estimates (37) and (38) we obtain (30), Q.E.D.

Proof of Theorem 4.3. From (24) and (30) we have

$$\|\hat{u}^N - \hat{u}\|^2 \leqslant ch(\|\hat{u}^N - \hat{u}\| + h) ,$$

which implies first order convergence of the control. Lemma 4.1 gives us an uniform estimate for the state.

4.2.2. Control constraints

The estimate obtained in Theorem 4.3 can be improved on the assumption that the state constraints are vacuous. Consider the problem to minimize the same functional (1) subject to the constraints (2) and (4). In order to shorten the presentation we shall use the same notation as in § 4.2.1, assuming that the state constraints are trivially satisfied, i.e. $\Theta(x,t) = -1$.

We assume that the hypotheses A1-A3 and one of the following conditions hold:

A4'. There exists a constant $\gamma > 0$ such that for all $t \in [0,1]$ and z

$$|G_c^T(t)z| \geq \gamma |z| ,$$

where $G_c(t)$ is defined as in A4.

A4''. The function $\Psi(u,t)$ does not depend on t.

The following theorem is to a certain extend a consequence of Theorem 4.1.

Theorem 4.4. There exist an optimal control $\hat{u}(\cdot)$,a corresponding trajectory $\hat{x}(\cdot)$, and optimal dual multipliers $q(\cdot),\lambda(\cdot)$,such that the relations (5)-(7) hold (with $\nu = 0$), and $\hat{x}(\cdot),\hat{u}(\cdot),\dot{q}(\cdot)$ are Lipschitz continuous on $[0,1]$ .If A4' holds then $\lambda(\cdot)$ is Lipschitz continuous on $[0,1]$.

In case A4'holds the proof of the Lipschitz continuity of $\hat{u}(\cdot)$ and $\lambda(\cdot)$ is identical to the proof, given in Hager [38] .If A4'' is satisfied,the Lipschitz continuity of $\hat{u}(\cdot)$ follows from the proof of Theorem 2.8.

The discrete problem associated with the one considered, consists in minimizing (8) subject to (9) and (11).Clearly, the statement of Theorem 4.2 holds with $\mu^N = 0$.

Let $q^N(t) = q_i^N$, $\hat{u}^N(t) = \hat{u}_i^N$ for $t \in [t_i,t_{i+1})$, $i = 0,\dots,N-1$.

Lemma 4.8.

$$\|q^N - q\|_C + \|\hat{u}^N - \hat{u}\|_C \leq c(\|\hat{u}^N - \hat{u}\| + h). \tag{39}$$

Proof. Lemmas 4.2 and 4.3, in which A4 is not used, imply that the sequences $q^N(\cdot)$ and $\hat{u}^N(\cdot)$ of functions are bounded uniformly on $[0,1]$ when $N \to \infty$.From (6) and (14) with $\nu = 0$ and $\mu^N = 0$ we get

$$|\Delta q_{i-1}| \leq (1+ch)|\Delta q_i| + ch(|\Delta x_i| + |\Delta u_i| + h),$$

$$|\Delta q_{N-1}| = O(h) .$$

Hence,applying Lemma 4.1

$$\|q^N - q\|_C \leq c(\|\hat{u}^N - \hat{u}\| + h) . \tag{40}$$

Let H(u,t) be the Hamiltonian for the continuous problem and $H_N(u,t_i)$ be defined as in Lemma 4.2.Then, applying Corollary 1.3 to the Maximum principle (see the proof of Theorem 2.7) we obtain

$$|(\frac{\partial}{\partial u}(H_N(\hat{u}_i^N,t_i) - H(\hat{u}_i^N,t_i))| \geq \alpha|\hat{u}_i^N - \hat{u}(t_i)| \quad ,i = 0,\ldots,N-1.$$

This inequality, combined with Lemma 4.1 and (40) gives us an uniform estimate for the control,Q.E.D.

<u>Theorem 4.5.</u>

$$\|\hat{u}^N - \hat{u}\|_C = O(h) .$$

Proof. Let $u_i^N = \hat{u}(t_i), i = 0,\ldots,N-1$, and x^N be determined by (9) for $u = u^N$.Then

$$\hat{I}_N \leq I_N(x^N,u^N) , \tag{41}$$

and

$$\begin{aligned}\hat{I}_N = {} & I_N(\hat{x}^N,\hat{u}^N) + \sum_{i=0}^{N-1} q^T(t_i)(\hat{x}_{i+1}^N - (I + hA_i)\hat{x}_i^N - hB_i\hat{u}_i^N) \\ \geq {} & I_N(x^N,u^N) + \sum_{i=0}^{N-1} h(\frac{\partial}{\partial u}f(\hat{x}(t_i),\hat{u}(t_i),t_i) - B_i^T q(t_i))^T(\hat{u}_i^N - u_i^N) \\ & + \sum_{i=0}^{N-1}(h(\frac{\partial}{\partial u}(f(x_i^N,u_i^N,t_i) - f(\hat{x}(t_i),u_i^N,t_i)))^T(\hat{u}_i^N - u_i^N) \\ & + (q(t_{i-1}) - (I + hA_i^T)q(t_i) + h\frac{\partial}{\partial x}f(x_i^N,u_i^N,t_i))^T(\hat{x}_i^N - x_i^N) \\ & + \alpha|\hat{u}_i^N - u_i^N|^2).\end{aligned} \tag{42}$$

The Maximum principle implies that the second term in the right hand side of the above inequality is nonnegative.It is easy to verify that

$$\max_{1\leq i\leq N}|x_i^N - \hat{x}(t_i)| = O(h) .$$

Hence, by the Lipschitz continuity of $\dot{q}(\cdot),\hat{u}(\cdot)$ and $\hat{x}(\cdot)$

$$\max_{1\leq i\leq N-1}|q(t_{i-1}) - (I + hA_i^T)q(t_i) + h\frac{\partial}{\partial x}f(x_i^N,u_i^N,t_i)| = O(h^2). \tag{43}$$

Finally, subtracting (41) from (42) and taking into account (43) we obtain

$$\|\hat{u}^N - \hat{u}\| = O(h).$$

Lemma 4.8 refines this estimate to the uniform metric,Q.E.D.

Remark 4.1. Clearly, the above estimate is exact for the Euler scheme considered.

Remark 4.2. The same proof can be used for a convex problem with a control constraining set V(t),which is closed and convex,provided that the optimal control is Lipschitz continuous.

4.2.3. Error estimates for the dual variables.

The technique, used in the previous two paragraphs enables us to estimate the convergence rate of the Lagrange multipliers.

Consider first the problem (1)-(4).Let $\lambda^N(\cdot)$, $\nu^N(\cdot)$, $p^N(\cdot)$, $q^N(\cdot)$ be step functions corresponding to λ^N, ν^N, p^N and q^N respectively.

Theorem 4.6. Suppose that A1 through A4 hold.Then

$$\|N\lambda^N - \lambda\| + |\nu^N - \nu| + \|p^N - p\| + |q^N - q|_C = O(h) .$$

Proof. Without loss of generality, let

$$Q^N = \Big(\sum_{i=0}^{N-1} ((\lambda_i^N)^{m_i})^2\Big)^{0.5} \neq 0$$

Since

$$\limsup_{N\to\infty} \max_{0\leq i\leq N-1} |\lambda_i^N| < +\infty$$

there exists $\beta_1 > 0$ such that

$$\psi(\hat{u}(t_i),t_i)_j + \beta_1\lambda_{ij}^N/Q^N \leqslant 0 \ , \ j\in m_i, \ i=0,\dots,N-1.$$

Hence

$$\sum_{i=0}^{N-1} \sum_{j\in m_i} \psi(\hat{u}(t_i),t_i)_j\lambda_{ij}^N \leqslant -\beta_1 Q^N .$$

In Lemma 4.5 we showed that

$$\sum_{i=0}^{N-1} \psi^T(\hat{u}(t_i),t_i)\lambda_i^N \geqslant -ch .$$

Thus

$$(Q^N)^2 = \sum_{i=0}^{N-1} |(\lambda_i^N)^{m_i}|^2 = O(h^2) .$$

In order to complete the proof one can apply this estimate to (33) and use Minkowski inequality,(37) and Theorem 4.3.

For the problem with vacuous state constraints we have $p(\cdot) = q(\cdot)$, and $p^N = q^N$.Moreover, the estimate for λ^N can be strengthened to the uniform metric.

Theorem 4.7. Suppose that A1-A3 and A4′hold for the problem (1)-(4) with trivial state constraints.Then

$$\| N\lambda^N - \lambda \|_C = O(h).$$

Proof.Using Theorem 4.5 and the proof of Lemma 4.4 we conclude that for N sufficiently large

$$(\lambda_i^N)^{m_i} = 0, \ i = 0,\dots,N-1 .$$

From (13) and A4′ we obtain consequently

$$\max_{0 \leq i \leq N-1} |(\lambda_i^N)^{c_i}| = O(h)$$

and

$$\gamma_0 |(\lambda_i^N)^{c_i} - h\lambda^{c_i}(t_i)| \leq h|B_i||q_i^N - q(t_i)| + |(C_i^N)^{c_i} - C^{c_i}(t_i)| \times$$
$$|(\lambda_i^N)^{c_i}| + h|\frac{\partial}{\partial u}(f(\hat{x}_i^N,\hat{u}_i^N,t_i) - f(\hat{x}(t_i),\hat{u}(t_i),t_i)| .$$

Taking advantage of Theorem 4.5 and lemmas 4.1 and 4.8 we complete the proof.

Remark 4.3. One can replace (10) by the following equivalent constraint

$$h\Psi(u_i,t_i) \leq 0 ,$$

which, obviously, does not change the discrete problem.In this case, however,the associate Lagrange multiplier will converge to the corresponding dual variable for the continuous problem.This conslusion may turn out to be important for the computational methods as primal

-dual methods and penalty function methods for solving constrained optimal control problems.

4.3. Integral constraints

4.3.1. Mixed inequality constraints

A discrete approximation to the following problem is examined: minimize the functional (1) subject to the constraints (2) and

$$\int_0^1 G(x(t),u(t),t)dt \leq 0 , \tag{44}$$

where $G\colon R^n \times R^m \times [0,1] \to R^q$,on the assumptions:

A5.The function $f(x,u,t)$ and the matrices $A(t)$,$B(t)$ satisfy A1 and A2.The function $G(\cdot)$ is C^2, $G(\cdot,t)$ is convex for each $t \in [0,1]$. There exist a continuous control $\bar{u}(\cdot)$ and a constant $\beta < 0$ such that

$$\int_0^1 G(\bar{x}(t),\bar{u}(t),t)_j dt \leq \beta$$

for all $j = 1,\ldots,q$, where $\bar{x}(\cdot)$ corresponds to $\bar{u}(\cdot)$.

By the classical arguments, see Joffe and Tikhomirov [42],p.317, and using the strong concavity of the Hamiltonian (see the proof of Theorem 2.8) we get

Theorem 4.8.There exist an optimal control $\hat{u}(\cdot)$,a corresponding trajectory $\hat{x}(\cdot)$ and optimal dual multipliers $p(\cdot)$, λ , $\lambda \geq 0$, $\lambda \in R^q$, such that $\hat{x}(\cdot)$,$\hat{u}(\cdot)$,$\dot{p}(\cdot)$ are Lipschitz continuous on $[0,1]$ and the following relations hold:

$$\frac{\partial}{\partial u} f(\hat{x}(t),\hat{u}(t),t) - B^T(t)p(t) - \frac{\partial}{\partial u} G^T(\hat{x}(t),\hat{u}(t),t)\lambda = 0, \tag{45}$$

$$\dot{p} = - A^T(t)p + \frac{\partial}{\partial x}(f(\hat{x}(t),\hat{u}(t),t) + G^T(\hat{x}(t),\hat{u}(t),t)\lambda) ,$$

$$p(1) = 0 , \tag{46}$$

$$\int_0^1 G^T(\hat{x}(t),\hat{u}(t),t)\lambda dt = 0. \tag{47}$$

The discrete problem, associated with the considered one consists in miminizing the functional (8) subject to (9) and

$$\sum_{i=0}^{N-1} hG(x_i, u_i, t_i) \leq 0 \ . \tag{48}$$

Introduce the Lagrange functional

$$L_N(x,u;p,\lambda) = I_N(x,u) + \sum_{i=0}^{N-1} (p_i^T(x_{i+1} - (I + hA_i)x_i - hB_iu_i)$$

$$+ hG^T(x_i, u_i, t_i)\lambda) \ , \tag{49}$$

where $p_i \in R^n$, $\lambda \in R^q$, $\lambda \geq 0$.By Kuhn-Tucher theorem, for large N we have:

<u>Theorem 4.9</u>. There exist unique solution $(\hat{x}^N, \hat{u}^N)$ and optimal dual multipliers p^N, λ^N such that

$$\hat{I}_N = I_N(\hat{x}^N, \hat{u}^N) = \min L_N(x,u;p^N,\lambda^N), \ x_o = x^o, \ x \in R^{Nn}, \ u \in R^{Nm} \ ,$$

$$\frac{\partial}{\partial u} f(\hat{x}_i^N, \hat{u}_i^N, t_i) - B_i^T p_i^N + \frac{\partial}{\partial u} G^T(\hat{x}_i^N, \hat{u}_i^N, t_i)\lambda^N = 0 \ , \tag{50}$$

$$p_{i-1}^N = (I + hA_i^T)p_i^N - h\frac{\partial}{\partial x}(f(\hat{x}_i^N, \hat{u}_i^N, t_i) + G^T(\hat{x}_i^N, \hat{u}_i^N, t_i)\lambda^N) \ ,$$

$$p_{N-1}^N = 0, \tag{51}$$

$$\sum_{i=0}^{N-1} hG^T(\hat{x}_i^N, \hat{u}_i^N, t_i)\lambda^N = 0 \ . \tag{52}$$

As in the previous section we assume that the discrete variables are step functions across the grid points.It is easy to verify that the statement of Lemma 4.2 holds and the sequence of λ^N is bounded when $N \to \infty$ .Moreover,using the proof of Lemma 4.3 we get

$$|\hat{u}^N - \hat{u}|^2 \leq ch(\|\hat{u}^N - \hat{u}\| + |\lambda^N - \lambda| + h) \ , \tag{53}$$

which gives us

$$|\hat{x}^N - \hat{x}|_C + \|\hat{u}^N - \hat{u}\| = O(\sqrt{h}) \ . \tag{54}$$

Observe that this result can be extended to a problem with an additional control constraints $u(t) \in V$, where V is a closed and convex set in R^m.

In order to improve the above estimate we apply the general scheme from Section 1.3.

Let $\mathcal{J}$ be the set of indices of the binding constraints for the continuous problem, that is

$$\mathcal{J} = \left\{ j \in \{1,\dots,q\} \; , \int_0^1 G(\hat{x}(t),\hat{u}(t),t)_j dt = 0 \right\} .$$

Then, from (54), using the proof of Lemma 1.2 we conclude that for N sufficiently large $\lambda_j^N = 0$, $j \in \{1,\dots,q\} \setminus \mathcal{J}$. Hence, we can limit our considerations to the case $\mathcal{J} = \{1,\dots,q\}$.

Let E(t,s) be the fundamental matrix solution of the equation $\dot{x} = A(t)x$. Introduce the matrices

$$\hat{G}_u(t) = \frac{\partial}{\partial u} G^T(\hat{x}(t),\hat{u}(t),t) \; , \quad \hat{G}_x(t) = \frac{\partial}{\partial x} G^T(\hat{x}(t),\hat{u}(t),t) \; ,$$

$$M_1 = \int_0^1 \hat{G}_u^T(t)\hat{G}_u(t)dt \; , \quad R(t) = \int_1^t E^T(t,s)\hat{G}_x(s)ds \; ,$$

$$K(t) = \hat{G}_u(t) - B^T(t)R(t) \; , \quad M_2 = \int_0^1 K^T(t)K(t)dt \; .$$

We assume that

A6. One of the following conditions holds:

(i) There exists $\gamma_0 > 0$ such that for every $z \in R^q$ and $t \in [0,1]$

$$|\hat{G}_u(t)z| \geq \gamma_0 |z| \; .$$

(ii) There exists γ_1 such that

$$\gamma_1 > \exp(\|A\|_C)\|B\|\|\hat{G}_x\|_{L^1} \quad \text{and} \quad z^T M_1 z \geq \gamma_1 |z|^2$$

for every $z \in R^q$.

(iii) The matrix M_2 is nonsingular.

Observe that if $G(\cdot)$ is independent of x, then (ii) is equivalent to (iii) and follows from (i).

Lemma 4.9.

$$|\lambda^N - \lambda| \leqslant c(\|\hat{u}^N - \hat{u}\| + h) . \tag{55}$$

Proof. Let $\Delta u = \hat{u}^N - \hat{u}$, $\Delta x = \hat{x}^N - \hat{x}$, $\Delta p = p^N - p$, $\Delta\lambda = \lambda^N - \lambda$. Let A6(i) hold. Then, from (45) and (50) we get

$$\gamma_0|\Delta\lambda| \leqslant c(|\Delta u(t_i)| + |\Delta x(t_i)| + |\Delta p(t_i)|) . \tag{56}$$

Subtracting (46) from (51) we obtain

$$|\Delta p(t_{i-1})| \leqslant (1 + h\|A\|_C)|\Delta p(t_i)| + h|\hat{G}_x(t_i)||\Delta\lambda|$$
$$+ ch(|\Delta x(t_i)| + |\Delta u(t_i)| + h) . \tag{57}$$

Substituting (56) in (57) and taking advantage of (16) and the discrete Gronwall lemma we get

$$\|\Delta p\|_C \leqslant c(\|\Delta u\| + h) .$$

This estimate combined with (56) gives us (55).

Now, let A6(ii) hold. Introduce the matrix

$$M_1^N = \sum_{i=0}^{N-1} h\hat{G}_u^T(t_i)\hat{G}_u(t_i) .$$

Since $M_1^N \to M_1$ when $N \to \infty$, for sufficiently small h and ε we have

$$(\Delta\lambda^T M_1 \Delta\lambda)^{0.5} \geqslant (\gamma_1 - \varepsilon)|\Delta\lambda| \geqslant \exp(\|A\|_C)\|B\|\|\hat{G}_x\|_{L^1}|\Delta\lambda| . \tag{58}$$

Furthermore, by (45) and (50)

$$(\gamma_1 - \varepsilon)|\Delta\lambda| \leqslant c(\|\Delta u\| + h) + \|B\|\|\Delta p\|_C . \tag{59}$$

Substituting (59) in (57) and taking into account (58) we obtain (55).

Finally, let A6(iii) hold. Let $E^N(t_i, s_j)$ be the fundamental matrix solution of the difference equation $x_{i+1} = (I + hA_i)x_i$. Denote

$$\hat{G}_u^N(t_i) = \frac{\partial}{\partial u} G(\hat{x}_i^N, \hat{u}_i^N, t_i)^T , \quad \hat{G}_x^N(t_i) = \frac{\partial}{\partial x} G(\hat{x}_i^N, \hat{u}_i^N, t_i)^T ,$$

$$R_i^N = \sum_{j=N-1}^{i} E^N(t_i,s_j)\hat{G}_x^N(s_j) \; , \quad K_i^N = \hat{G}_u^N(t_i) - B_i^T R_i^N \; ,$$

and let

$$q(t) = R(t)\lambda - p(t) \; , \; q_i^N = R_i^N \lambda^N - p_i^N \; .$$

Then from (45),(46),(50) and (51) we get the following optimality relations

$$\frac{\partial}{\partial u} f(\hat{x}(t),\hat{u}(t),t) + B^T(t)q(t) + K(t)\lambda = 0 \; , \tag{60}$$

$$\dot{q} = - A^T(t)q - \frac{\partial}{\partial x} f(\hat{x}(t),\hat{u}(t),t) \; , \quad q(1) = 0 \; , \tag{61}$$

$$\frac{\partial}{\partial u} f(\hat{x}_i^N,\hat{u}_i^N,t_i) + B_i^T q_i^N + K_i^N \lambda^N = 0 \; , \tag{62}$$

$$q_{i-1}^N = (I + hA_i^T)q_i^N + h\frac{\partial}{\partial x} f(\hat{x}_i^N,\hat{u}_i^N,t_i) - h^2 A_i^T \hat{G}_x^N(t_i)\lambda^N \; ,$$

$$q_{N-1}^N = h\hat{G}_x^N(t_{N-1})\lambda^N \; . \tag{63}$$

Subtracting (60) from (62) and (61) from (63) and repeating the argument of the latter case we get (55),Q.E.D.

Theorem 4.10. The following estimation holds

$$\|\hat{x}^N - \hat{x}\|_C + \|\hat{u}^N - \hat{u}\|_C = O(h) \; .$$

Proof.Substituting (55) in (53) we immediately obtain

$$\|\hat{u}^N - \hat{u}\| = O(h) \; .$$

Then

$$\|\hat{x}^N - \hat{x}\|_C + \|p^N - p\|_C + |\lambda^N - \lambda| = O(h) \; .$$

Thus, by (45) and (50) as in the proof of Lemma 4.8 we obtain the desired estimate.

4.3.2.An example with equality constraints.

The analysis, presented in the previous paragraph, can be applied to problems with integral equality constraints.A classical example of this type is the fixed final state problem: minimize (1) for the system (2) when $x(1) = x^1$.As it was demonstrated in § 2.3.5 , this problem can be regarder as a problem with integral control constraints

$$\int_0^1 P(t)u(t)dt = x^1 - E(1,0)x^o , \qquad (64)$$

where $P(t) = E(1,t)B(t)$.The corresponding discrete problem consists in minimizing (8) for the system (9) when

$$\sum_{i=0}^{N-1} hP_i^N u_i = x^1 - E^N(1,0)x^o , \qquad (65)$$

where $P_i^N = E^N(1,t_i)B_i$.We assume that

A7. The system (2) is controllable on $[0,1]$, that is, the matrix

$$M = \int_0^1 P(t)P^T(t)dt$$

is nonsingular.

Define the controllability matrix for the discrete system

$$M^N = \sum_{i=0}^{N-1} hP_i^N(P_i^N)^T .$$

Clearly, for N sufficiently large the matrix M^N is nonsingular as well.Observe that M corresponds to the matrices M_1 and M_2 in A6. Then, in order to get the estimate (55) for the Lagrange multipliers λ and λ^N , associated with the constraints (64) and (65), it is sufficient to show that the sequence $|\lambda^N|$ is bounded.

Define the control

$$\bar{u}_i^N = (P_i^N)^T(M^N)^{-1}(- \lambda^N/|\lambda^N| + x^1 - E^N(1,0)x^o)$$

for $i = 0,\dots,N-1$ and for large N.Then $\bar{u}^N$ is bounded uniformly in $[0,1]$ together with the corresponding discrete trajectory $\bar{x}^N$.Then, by the duality relation

$$\hat{I}_N \leqslant I_N(\bar{x}^N,\bar{u}^N) + (\lambda^N)^T(\sum_{i=0}^{N-1} hP_i^N\bar{u}_i^N - x^1 + E^N(1,0)x^o)$$

we get

$$-|\lambda^N| \leqslant \hat{I}_N - I_N(\bar{x}^N,\bar{u}^N),$$

hence, the sequence of λ^N is bounded.Parallelly to the proof of Theorem 4.10 we obtain

$$\|\hat{x}^N - \hat{x}\|_C + \|\hat{u}^N - \hat{u}\|_C = O(h) .$$

Concerning high-order integration schemes, there are no convergence estimations known to the author.Actually, the main difficulty in such an analysis would stem from the fact that the optimal control of constrained problems has, as a rule, discontinuous derivative at the points where the binding constraints change.Therefore, on order to get exact error bounds of, say Runge-Kutta schemes,one has to know or estimate these "contact" points.In this direction the method, proposed in Hager [36] for dual approximations, which uses splines with free knots in order to approximate both the optimal control and the contact points, could be very helpful.

CHAPTER 5

SENSITIVITY ANALYSIS OF THE OPEN - LOOP CONTROL STRUCTURE WITH CONSTRAINED CONTROLS

5.1. Real sensitivity analysis

The major part of this work has been concerned with determining the optimal performance and control variations under known changes of the model. Indeed, this is a fundamental problem in the sensitivity analysis of optimization problems. From a practical point of view, however, it is often more important to determine the quantitative effects in the pair "model - real process", that is, in a control system which consists of a model applied to a real process in a given optimization structure. The basic concept in this direction, called real sensitivity analysis, was proposed by Wierzbicki [77]. Here we briefly outline the real sensitivity problem.

Consider the following optimization problem

$$Q(x,u,a) \longrightarrow \inf_{(x,u)} \qquad (1)$$

subject to

$$P(x,u,a) = 0 \ , \qquad (2)$$

$$u \in U(a) \ , \qquad (3)$$

where $a \in A \subset B_a$, $U(a) \subset B_u$, $Q(\cdot,a)\colon B_x \times U(a) \to R^1$, $P\colon B_x \times B_u \times A \to B_p$, B_x, B_u, B_p and B_a are suitable spaces. This problem can be regarded as an optimal control problem with a state x, a control u and a parameter a provided that the constraining relation (2) determines uniquely x for any given u and a and thus can be interpreted as a state equation. The set U(a) represents the admissible set of controls.

In real applications, the determination of an optimal control is based on a mathematical model,e.g.(1) - (3), for a given value of the parameter a.Obviously, this <u>basic model</u> may differ from the reality.One can assume that the reality is represented by a model which differs from the basic one only in the value of the parameter, that is

$$Q(x,u,\alpha) \longrightarrow \inf_{(x,u)} , \tag{4}$$

$$P(x,u,\alpha) = 0 , \tag{5}$$

$$u \in U(\alpha) , \tag{6}$$

for some $\alpha \in A$.In fact, such an assumption is quite general.For example, the change from α to a may represent a reduction of a partial differential equation to an ordinary one, a finite- difference approximation etc.The model (4) - (6) is called <u>extended model</u> of the problem considered.

The optimal control can be applied to the real process (the extended model) in the same way as it is determined, that is in the so-called <u>open-loop structure</u> ,see Fig. 5.1.

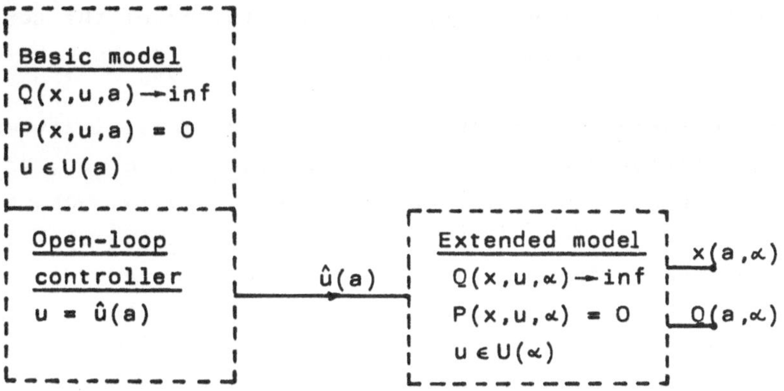

Fig. 5.1.

Let $\hat{u}(a)$ be the optimal control for the basic model and $\hat{Q}(\alpha)$ be the optimal value for the extended model.The state trajectory $x(a,\alpha)$ of the extended model, which results from $\hat{u}(a)$,satisfies the equation

$$P(x,\hat{u}(a),\alpha) = 0 .$$

The set $U(a)$ of addmissible controls in the basic model represents the possibilities of the controller.The set $U(\alpha)$ in the extended model may be interpreted as a set of theoretically admissible controls. It is natural to assume that $U(a) \subset U(\alpha)$.Then, since $\hat{u}(a) \in U(\alpha)$, the functional

$$Q(a,\alpha) = Q(x(a,\alpha),\hat{u}(a),\alpha)$$

is well-defined and represents the (nonoptimal) performance of the open-loop structure.

The functional

$$S(a,\alpha) = Q(a,\alpha) - \hat{Q}(\alpha)$$

can be interpreted as the performance loss due to an imperfect knowledge (or a simplification) of the reality.This functional is called sensitivity measure of the optimal control system in the open-loop structure.It is the primary purpose of the real sensitivity analysis to investigate the properties of the sensitivity measure.

Besides this basic definition, sensitivity measures can be introduced for every control structure.Here we limit our considerations to the open-loop structure.A complete presentation of the concepts of the real sensitivity analysis can be found in the book by Wierzbicki [78] .

The straightforward approach to estimate the sensitivity measure consists in a Taylor expansion about a or α.Such a method, based on some symmetry properties of the derivatives of $S(a,\alpha)$, was proposed in Wierzbicki and Dontchev [79] , see also Wierzbicki [78].This method, however, requires differentiability of the optimal control $\hat{u}(a)$ with respect to the perturbation parameter.This means that it can not be applied to problems with inequality constraints, to the case when the basic model is a discrete approximation of the extended one etc. Developing the ideas of Chapter 1,in this chapter we propose a new scheme for estimating the sensitivity measure of convex problems.The general estimate,obtained in Section 5.2,is applied to various perturbed control problems.Section 5.3 deals with a regularly perturbed model with control constraints.Section 5.4 considers the validity of the order reduction.Section 5.5 is devoted to the case when the perturbation is represented by a finite-difference approximation.Section 5.6 contains sensitivity evaluations of a control system described by a partial differential equation.

5.2. Basic estimates.

Let B_x, B_u, B_p and B_a be Banach spaces. Consider the problem (1)-(3) on the assumptions:

A1. For every $a \in A$ the set $U(a)$ is convex, the problem (1)-(3) has a solution $(\hat{x}(a), \hat{u}(a))$ and for every $u \in U(a)$ there exists $x(u,a)$ such that $P(x(u,a),u,a) = 0$. For every $a \in A$ the functional

$$J(u,a) = Q(x(u,a),u,a)$$

is convex and Frechet differentiable with respect to u with continuous derivative $J'(\cdot,a)$ on $U(a)$.

Then sensitivity measure can be expressed as

$$S(a,\alpha) = J(\hat{u}(a),\alpha) - J(\hat{u}(\alpha),\alpha) ,$$

where a and α are given elements of the set A.

Let $\|\cdot\|$ denote the norms of all the considered spaces and $\langle\cdot,\cdot\rangle$ be the duality.

Proposition 5.1. For every $a,\alpha \in A$ such that $\hat{u}(a) \in U(\alpha)$ the following inequality holds

$$S(a,\alpha) \leqslant \langle J'(\hat{u}(a),\alpha) - J'(\hat{u}(a),a),\ \hat{u}(a) - \hat{u}(\alpha)\rangle$$

$$+ \|J'(\hat{u}(a),a)\| \inf_{u \in U(a)} \|u - \hat{u}(\alpha)\| . \tag{8}$$

Proof. The convexity of $J(\cdot,a)$ yields

$$S(a,\alpha) \leqslant \langle J'(\hat{u}(a),\alpha), \hat{u}(a) - \hat{u}(\alpha)\rangle \tag{9}$$

and

$$\langle J'(\hat{u}(a),a), u - \hat{u}(a)\rangle \geqslant 0$$

for every $u \in U(a)$. Adding the last expression to the right-hand side of (9) and using Schwarz inequality we obtain (8), Q.E.D.

Notice that if $\hat{u}(a) \in \operatorname{int} U(a)$ or $\hat{u}(\alpha) \in U(a)$, then the second term in (8) vanishes.

In order to improve (8) we assume that:

A2. For every $a \in A$ the functional $J(\cdot,a)$ is strongly convex on $U(a)$ with a constant $\varkappa$ independent of a.

Proposition 5.2. For every $a,\alpha\in A$ such that $\hat{u}(a)\in U(\alpha)$ the following estimation holds

$$S(a,\alpha) \leqslant V_1^2 + V_1\sqrt{V_2} + 2\varkappa V_2 , \tag{10}$$

where

$$V_1 = \| J'(\hat{u}(a),\alpha) - J'(\hat{u}(a),a)\| /\sqrt{2\varkappa} , \tag{11}$$

$$V_2 = \| J'(\hat{u}(a),a)\| \inf_{u\in U(a)} \|u - \hat{u}(\alpha)\|/2\varkappa . \tag{12}$$

The proof follows from Proposition 1.2 in Chapter 1, applied to (8).

Corollary 5.1. Let $\hat{u}(a)\in \operatorname{int} U(a)$ or $\hat{u}(\alpha)\in U(a)$.Then

$$S(a,\alpha) \leqslant \|J'(\hat{u}(a),\alpha) - J'(\hat{u}(a),a)\|^2/2\varkappa . \tag{13}$$

Moreover, if $J'(\hat{u}(a),\cdot)$ is continuously differentiable on A, then

$$S(a,\alpha) \leqslant (\| \frac{\partial}{\partial a} J'(\hat{u}(a),a)\|^2/2\varkappa)\|a - \alpha\|^2 + o(\|a - \alpha\|^2).\tag{14}$$

Example 5.1. Let $B_a = B_u = U(a) = R^1$ and

$$J(u,a) = 0.5u^2 - au .$$

Then $\varkappa = 0.5$, $J'(\hat{u}(a),\alpha) = a - \alpha$, $J'(\hat{u}(\alpha),\alpha) = 0$, $J(\hat{u}(\alpha),\alpha) = - 0.5\alpha^2$, $J(\hat{u}(a),\alpha) = 0.5a^2 - a\alpha$ and

$$S(a,\alpha) = 0.5(a-\alpha)^2 = \|J'(\hat{u}(a),\alpha) - J'(\hat{u}(a),a)\|^2 ,$$

that is (13) holds as equality.

Remark 5.1. Since $S(a,\alpha) \geqslant 0$ and $S(\alpha,\alpha) = 0$ then if $S(\cdot,\alpha)\in C^2$, we have $S_a'(\alpha,\alpha) = 0$ and the deviation of $S(a,\alpha)$ from zero will be determined by the second term in the Taylor expansion.Observe that the same convergence rate follows from (14) without any differentiability assumption for $S(\cdot)$.

Remark 5.2. Actually, it is not necessary to assume that $\hat{u}(a)$ is an element of $U(\alpha)$.In this case, by Proposition 2.1 we obtain a next term in (10) representing the destance between $\hat{u}(a)$ and $U(\alpha)$.This means, however, that the optimal control of the basic model may not satisfy the constraints of the real process.For more detailed discussion of ill-posed sensitivity problems, see Wierzbicki [78].

Concerning other control structures, in order to apply our analysis we need some convexity and differentiability conditions for the optimal control law.Such conditions would be not natural even when the the simplest closed-loop structure with constrained controls is considered, see Brunovský [6].We shall not go into this further noting only that the sensitivity evaluations for constrained problems in, say, the closed-loop structure, will depend essentially on the concrete form of the optimal controller.

5.3.Regular perturbations

Consider the problem studied in § 2.3.4:

$$J_\varepsilon(u(\cdot)) = 0.5 \int_0^1 (|x(t)|^2 + |u(t)|^2)dt \longrightarrow \inf$$

$$\dot{x} = A_\varepsilon(t)x + B(t)u, \quad t\in[0,1],\ x(0) = x^0 ,$$

$$u\in U_\varepsilon = \{u(\cdot)\in L^2(R^m),\ u(t)\in V_\varepsilon \quad \text{for a.e.}\ t\in[0,1]\},$$

on the same assumptions.Let $V_0 \subset V_\varepsilon$ for all $\varepsilon\in[0,\varepsilon_0]$.The basic model corresponds to $\varepsilon = 0$.The sensitivity measure will have the form

$$S(\varepsilon) = J_\varepsilon(\hat{u}_0(\cdot)) - J_\varepsilon(\hat{u}_\varepsilon(\cdot)) .$$

Using Theorem 2.8, from (10)-(12) one can deduce

<u>Theorem 5.1</u>. For every $\varepsilon\in[0,\varepsilon_0]$ the following estimation holds

$$S(\varepsilon) \leqslant k_1\varepsilon^2 + k_1\varepsilon\sqrt{k_2 d(\varepsilon)} + k_2 d(\varepsilon) ,$$

where

$$d(\varepsilon) = \min_{u\in V_0}\ \max_{0\leqslant t\leqslant 1} |u - \hat{u}_\varepsilon(t)| ,$$

$$k_1 = \|B\|^2 e^{2a} L_a^2 (e^a \|\hat{x}_0\|_{L^1} + \|p_0\|_{L^1})^2 ,$$

$$k_2 = \|\hat{u}_0 - B^T p_0\| .$$

and, as in Thorem 2.8, $a = \sup \|A_\varepsilon\|_C$ and L_a is the Lipschitz constan of the matrix $A_\varepsilon(t)$, $p_0(\cdot)$ is the optimal adjoint variable.

Example 5.2.

$$J_\varepsilon(u(\cdot)) = 0.5((x(1) - 1)^2 + \int_0^1 u(t)^2 dt) \longrightarrow \inf ,$$

$$\dot{x} = \varepsilon u,\ x(0) = 0,\quad 0 \leq u(t) \leq 1-t \ .$$

The optimal control is

$$\hat{u}_\varepsilon(t) = \begin{cases} \varepsilon/(1+\varepsilon^2) & \text{for } 0 \leq t \leq 1-\varepsilon/(1+\varepsilon^2) \ , \\ 1 - t & \text{for } 1-\varepsilon/(1+\varepsilon^2) < t \leq 1 \ . \end{cases}$$

Clearly, although the control constraining set is time varying and the functional is depending explicitely on the final state, the estimate (10) applies as well.The derivative $J'_\varepsilon(\cdot)$ is given by

$$(J'_\varepsilon(u(\cdot)))(t) = u(t) - \varepsilon\, p(t) \ ,$$

where

$$\dot{p} = 0 \ ,\ p(1) = 1-x(1) \ ,\ \dot{x} = \varepsilon u \ ,\ x(0) = 0.$$

Hence

$$J'_\varepsilon(\hat{u}_0(\cdot)) = -\varepsilon \ , \qquad J'_0(\hat{u}_0(\cdot)) = 0 \ ,$$

and, from (13)

$$S(\varepsilon) \leq \varepsilon^2 \ .$$

5.4. Singular perturbations

We take up the problem from § 3.5.2, namely

$$J_\beta(u(\cdot)) = g(x(1)) + \int_0^1 (f(x(t),y(t),t) + h(u(t),t))dt \longrightarrow \inf$$

subject to

$$\dot{x} = A_1(t)x + A_2(t)y + B_1(t)u \;,\; x(0) = x^o \;,$$

$$\beta\dot{y} = A_3(t)x + A_4(t)y + B_2(t)u \;,\; y(0) = y^o \;,$$

$$u(\cdot) \in U = \{u(\cdot) \in L^2(R^r),\; u(t) \in V \;\text{ for a.e. } t \in [0,1]\} .$$

The low-order system

$$\dot{x} = A_o(t)x + B_o(t)u \quad,\; x(0) = x^o \;,$$

$$0 = A_3(t)x + A_4(t)y + B_2(t)u$$

represents the basic model.We apply the optimal control $\hat{u}_o(\cdot)$ of the limit problem to the full-order system.Then the performance loss is given by the sensitivity measure

$$S(\beta) = J_\beta(\hat{u}_o(\cdot)) - J_\beta(\hat{u}_\beta(\cdot)).$$

Using the proof of Theorem 3.8 and (13) we get

Theorem 5.2. For every $\beta \in (0,\beta_1)$

$$S(\beta) \leq r_1\beta + r_2\beta^2 \;,$$

where

$$r_1 = (\|B_1\| d_4 + \|B_2\| d_6)^2/2\varkappa \;,$$

$$r_2 = (\|B_1\| d_5 + \|B_2\| d_7)^2/2\varkappa \;,$$

and the constants $d_4 \div d_7$ are determined in the proof of Theorem 3.8. If the assumptions of Theorem 3.9 hold then $r_1 = 0$.

Example 5.3.

$$J_\beta(u(\cdot)) = \int_0^1 (0.5u(t)^2 + y(t))dt \longrightarrow \min \;,$$

$$\beta\dot{y} = -y + u \;,\; y(0) = 0 \;.$$

The optimal controls for the perturbed and for the limit problems are

$$\hat{u}_\beta(t) = -1 + \exp((t-1)/\beta) \quad,\quad \hat{u}_o(t) = -1.$$

Furthermore

$$(J'_\beta(\hat{u}_o(\cdot))(t) = \hat{u}_o(t) - p(t) ,$$

where

$$\beta\dot{p} = p + 1 \quad , \; p(1) = 0 .$$

Hence

$$(J'_\beta(\hat{u}_o(\cdot))(t) = \exp((t-1)/\beta) .$$

From (13) we have

$$S(\beta) \leqslant 0.5\beta + o(\beta) .$$

By elementary but tedious computations one can obtain

$$S(\beta) = \beta/4 + o(\beta) .$$

Thus, our method gives the exact convergence rate, but the constant may be not exact.In this case $r_1 \neq 0$.This follows from the fact that the fast states are included in the performance index, compare with Dontchev [12] .

The singular perturbation analysis can be also applied to problems with a large optimization horizon.Consider the problem

$$J_T(u(\cdot)) = \int_0^T (f(y(t)) + h(u(t)))dt \longrightarrow \inf$$

$$\dot{y} = A_4 y + B_2 u \quad , \; y(0) = y^o ,$$

$$u(t) \in V \text{ for a.e. } t \in [0,T] ,$$

where $T \gg 1$ and $f(\cdot),h(\cdot),A_4,B_2$ are as in § 3.5.2.Changing the time-scale $s = t/T$ and letting $\beta = 1/T$ we get a singularly perturbed problem without slow states.Suppose that the system is controlled by the solution of the mathematical programming problem obtained for $\beta = 0$ $(T = +\infty)$:

$$f(A_4^{-1}B_2u) + h(u) \longrightarrow \inf \quad , \quad u \in V .$$

From Theorem 5.2 we conclude that the performance loss due to neglecting the dynamics of the process is

$$S(T) = O(1/T) .$$

Notice that this result becomes nontrivial only when the (fast) states are involved in the performance index.

5.5. Finite-difference approximation

Consider the following situation:Suppose that a continuous process is to be controlled optimally but we can solve only its finite-difference approximation.Let the discrete optimal control, derived by the finite-dimensional model, be a step function across the grid points.When applied to the continuous process, the discrete control results in a deviation from the optimality, represented by the sensitivity measure.

Let us take the problem from § 4.2.2, namely

$$I(x(\cdot),u(\cdot)) = \int_0^1 f(x(t),u(t),t)dt \longrightarrow \inf$$

$$\dot{x} = A(t)x + B(t)u \quad , \; x(0) = x^o ,$$

$$\Psi(u(t),t) \leqslant 0 .$$

Denote by $\hat{u}^N(\cdot)$ the discrete optimal control, solving the corresponding approximate problem

$$I_N(x,u) = \sum_{i=0}^{N-1} hf(x_i,u_i,t_i) \longrightarrow \inf$$

$$x_{i+1} = (I + hA_i)x_i + hB_iu_i \quad , \; x_o = x^o ,$$

$$\Psi(u_i,t_i) \leqslant 0 \quad , \; i = 0,\dots,N-1, \; h = 1/N .$$

Let $\tilde{x}^N(\cdot)$ be the continuous system responce to $\hat{u}^N(\cdot)$.Then the sensitivity measure will have the form

$$S(h) = I(\tilde{x}^N(\cdot),u^N(\cdot)) - I(\hat{x}(\cdot),\hat{u}(\cdot)) ,$$

where $(\hat{x}(\cdot),\hat{u}(\cdot))$ is the solution of the continuous problem.

Denote by $H(u,t)$ the Hamiltonian, that is

$$H(u,t) = - f(\hat{x}(t),u,t) + p^T(t)B(t)u ,$$

where $p(\cdot)$ is the optimal adjoint variable.

Theorem 5.3. There exists a constant c such that for N sufficiently large the following estimation holds

$$0 \leqslant \langle \frac{\partial}{\partial u} H(\hat{u}),\hat{u} - \hat{u}^N \rangle \leqslant S(h) \leqslant c(h^2 + h\|\frac{\partial}{\partial u} H(\hat{u})\|) . \qquad (15)$$

Proof.In order to apply the general estimates from Section 5.2 one should use an embedding of the discrete problem in the class of the continuous one.Here we present a direct, more simple proof.

Using the optimality conditions, see Theorem 4.1, we have

$$\begin{aligned} S(h) &= I(\tilde{x}^N(\cdot),\hat{u}^N(\cdot)) + \langle p,\dot{\tilde{x}}^N - A\tilde{x}^N - B\hat{u}^N\rangle \\ &\quad - I(\hat{x}(\cdot),\hat{u}(\cdot)) + \langle p,\dot{\hat{x}} - A\hat{x} - B\hat{u} \rangle \\ &\leqslant \langle \dot{p} + A^T p - \frac{\partial}{\partial x} f(\tilde{x}^N,\hat{u}^N),\hat{x} - \tilde{x}^N \rangle \\ &\quad + \langle \frac{\partial}{\partial u} f(\tilde{x}^N,\hat{u}^N) - B^T p,\hat{u}^N - \hat{u} \rangle \\ &\leqslant c(\|\tilde{x}^N - \hat{x}\|^2 + \|\hat{u}^N - \hat{u}\|^2 + \|\frac{\partial}{\partial u} H(\hat{u})\| \|\hat{u}^N - \hat{u}\|). \end{aligned}$$

Theorem 4.5 and the Gronwall lemma yield

$$\|\tilde{x}^N - \hat{x}\| + \|\hat{u}^N - \hat{u}\| = O(h) .$$

The last two relations gives us the upper bound in (15).The lower estimate follows from the relations

$$\begin{aligned} S(h) &= I(\tilde{x}^N(\cdot),\hat{u}^N(\cdot)) + \langle p,\dot{\tilde{x}}^N - A\tilde{x}^N - B\hat{u}^N\rangle \\ &\quad - I(\hat{x}(\cdot),\hat{u}(\cdot)) + \langle p,\dot{\hat{x}} - A\hat{x} - B\hat{u} \rangle \\ &\geqslant \langle \dot{p} + A^T p - \frac{\partial}{\partial x} f(\hat{x},\hat{u}),\hat{x} - \hat{x}^N \rangle \\ &\quad + \langle \frac{\partial}{\partial u} H(\hat{u}),\hat{u} - \hat{u}^N \rangle = \langle \frac{\partial}{\partial u} H(u),\hat{u} - \hat{u}^N \rangle \text{ ,Q.E.D.} \end{aligned}$$

Clearly, since the distance between $\hat{u}^N(\cdot)$ and $\hat{u}(\cdot)$ is exactly $O(h)$, the expected convergence rate of $S(h)$ for constrained problems will be $O(h)$.When the constraints are vacuous, the convergence rate increases twice.

We present two examples which show that, slightly changing the problem, one can essentially change the convergence rate of the sensitivity measure.

Example 5.4.

$$\int_0^1 (u(t) - t)^2 dt \longrightarrow \inf$$

$$u(t) \leq 0.5 \; .$$

The optimal control is

$$\hat{u}(t) = \begin{cases} t & \text{for } 0 \leq t \leq 0.5 \; , \\ 0.5 & \text{for } 0.5 < t \leq 1 \; . \end{cases}$$

In the interval $[0.5, 1]$, where the constraints are binding, we have $\hat{u}^N(t) = \hat{u}(t)$.Hence, by (15) we get $S(h) = O(h^2)$.

Example 5.5.

$$\int_0^1 (u(t) - 0,5)^2 dt \longrightarrow \inf$$

$$u(t) \leq t \quad .$$

The solution is the same, but $\|\hat{u}^N - \hat{u}\| = O(h)$ when $\frac{\partial}{\partial u} H(\hat{u}(t),t) < 0$, hence $S(h) = O(h)$.

5.6.Sensitivity analysis of a system described by a parabolic equation

As an application of the real sensitivity analysis we consider the following problem: for fixed $\underline{a}$ minimize the functional

$$J(u(\cdot),\underline{a}) = 0.5(k \int_0^1 u(t)^2 dt + \int_0^1 (p(x,1) - p_1)^2 dx) \qquad (16)$$

subject to the constraints

$$\frac{\partial}{\partial t} p(x,t) = d \frac{\partial^2}{\partial x^2} p(x,t) \quad ,(x,t) \in (0,1)\times(0,1] \, , \quad (17)$$

$$p(0,t) = c \quad , x \in (0,1) \, , \quad (18)$$

$$\frac{\partial}{\partial x} p(0,t) = - b(u(t) - p(0,t)) \, , \; t \in (0,1] \, , \quad (19)$$

$$\frac{\partial}{\partial x} p(1,t) = -f(t) \quad ,t \in (0,1] \, , \quad (20)$$

$$u(\cdot) \in U(a) = \left\{ u(\cdot) \in L^2(R^1), \; u(t) \in [0,u_m] \; \text{ for a.e.} t \in [0,1] \right\} ,$$

where k,d,c,b,u_m are constants, $f(\cdot) \in C^1(R^1)$. Here $\underline{a}$ represents the data of the problem. The solution of the equation (17) satisfies (18) and (20) in the usual sense, and (19) in the weak sense, see Vasilev [71].

This problem is classically known as a model for optimal heating of a metal bar with uniform thickness, see Rolewicz [66]. An other, less popular but not less reasonable interpretation of the problem is concerned with the gas transmission process, see the report [58]. Consider a segment of a gas-pipe with a pump, connected with the source, and with a consumer. The function $p(\cdot)$ represents the presure along the gas-pipe, the control $u(\cdot)$ corresponds to the power of the pump. The first term in the performance index (16) is related to the costs of the gas-transmission, the second one expresses the requirement for the final presure. The parameter d represents the technological characteristics of the gas-pipe, e.g. diameter, lenght; b is connected with the parameters of the pump.

By a classical argument, see e.g. Lions [49], there exists unique solution of the problem considered. We found the optimal control numerically, see Fig. 5.2, for the following values of the parameters: $d = 0.8, b = 1., c_0 = 3., k = 0,2, p_1 = 8., f_0(t) = 0.5 + 0.1\sin 10\pi t$, $u_m = 7.85$.

We analyze the sensitivity of the open-loop structure with respect to the parameters: $d,b,c,f(t),p_1,u_m$, assuming that the deviation of the function $f(\cdot)$ is given by

$$f(t) = f_0(t) + \varepsilon_0 + \varepsilon \sin 10\pi t \quad ,$$

where ε_0 and ε are zeros in the basic model. The upper bound of the control in the extended model is $u_m + \delta u$, $\delta u \geqslant 0$. In other words, the

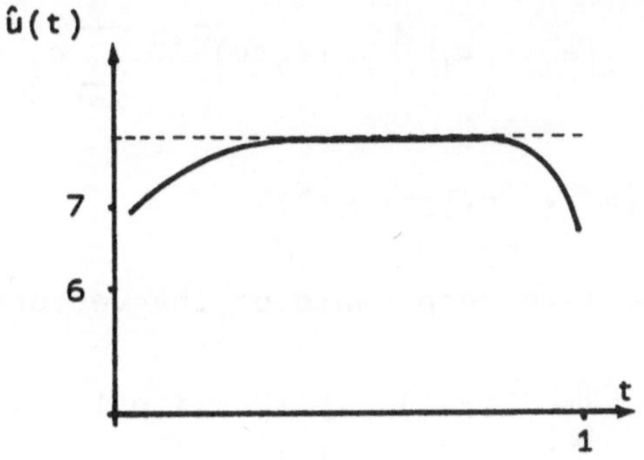

Fig.5.2

parameters in the extended model are represented by the vector

$$\underline{\alpha} = (d, b, c, \varepsilon_o, \varepsilon, p_1, u_m) \ ,$$

which in the basic model is

$$\underline{a} = (0.8, 1., 3., 0, 0, 8., 7.85).$$

In order to estimate the sensitivity measure we use the following standard result:

Lemma 5.1.The functional $J(\cdot, \underline{a})$ is Frechet differentiable in $L^2(R^1)$ and its derivative is given by

$$(J'(u(\cdot), \underline{a}))(t) = -\ dbq(0,t) + ku(t) \ ,$$

where $q(\cdot)$ solves the adjoint equation

$$\frac{\partial}{\partial t} q(x,t) = -\ d \frac{\partial^2}{\partial x^2} q(x,t) \ , \quad (x,t) \in (0,1) \times (0,1] \ ,$$

$$\frac{\partial}{\partial x} q(0,t) = bq(0,t) \ , \qquad t \in (0,1) \ ,$$

$$\frac{\partial}{\partial x} q(1,t) = 0 \ , \qquad t \in (0,1) \ ,$$

$$q(x,1) = p(x,1) - p_1 \ , \quad x \in (0,1) \ ,$$

This lemma, applied to Corollary 5.1 gives the following estimation for the sensitivity measure :

$$S(\underline{a},\underline{\alpha}) \leqslant \sum_{i=1}^{6} c_i |a_i - \alpha_i|^2 + (c_7 \delta u)^{0.5} \sum_{i=1}^{6} c_i |a_i - \alpha_i|$$

$$+ k c_7 \delta u \quad + \quad o(|\underline{a} - \underline{\alpha}|^2),$$

where a_i and α_i are the i-th components of the vectors $\underline{a}$ and $\underline{\alpha}$ and

$$c_i = \left\| \frac{\partial}{\partial a_i} J'(\hat{u}(\cdot),\underline{a}) \right\|^2 / k \quad ,i = 1,\ldots,6 \ ,$$

$$c_7 = \quad \| J'(\hat{u}(\cdot),\underline{a}) \| / k \quad ,$$

where $\hat{u}(\cdot)$ is the optimal control for the basic model.

Let $\hat{p}(\cdot)$ be the optimal state for the basic model and $\hat{q}(\cdot)$ be the corresponding solution of the adjoint equation.Then

$$c_1 = \quad \| dbq_d(0,\cdot) + bq(0,\cdot) \|^2 / k,$$

where $q_d(\cdot)$ is the solution of

$$\frac{\partial}{\partial t} q(x,t) = - d \frac{\partial^2}{\partial x^2} q(x,t) - \frac{\partial^2}{\partial x^2} \hat{q}(x,t) \ ,$$

$$(x,t) \in (0,1)\times(0,1] \ ,$$

$$\frac{\partial}{\partial x} q(0,t) = bq(0,t) \ , \quad t \in (0,1) \ ,$$

$$\frac{\partial}{\partial x} q(1,t) = 0 \ , \qquad t \in (0,1) \ ,$$

$$q(x,1) = p_d(x,1) \ , \quad x \in (0,1) \ ,$$

while $p_d(\cdot)$ solves

$$\frac{\partial}{\partial t} p(x,t) = d \frac{\partial^2}{\partial x^2} p(x,t) + \frac{\partial^2}{\partial x^2} \hat{p}(x,t) \ ,$$

$$(x,t) \in (0,1)\times(0,1]$$

$$\frac{\partial}{\partial x} p(0,t) = bp(0,t) \ , \ t \in (0,1] \ ,$$

$$\frac{\partial}{\partial x} p(1,t) = 0 \ , \quad t \in (0,1] \ ,$$

$$p(x,0) = 0 \ , \ x \in (0,1) \ .$$

Similarly, one can derive the constants $c_2, \ldots, c_7$. The sensitivity constants, obtained for $k = 0.1$, 0.2 and 0.3 are given in the table below. In the case $k = 0.3$ the control constraints are not binding, therefore $c_7 = 0$.

k	c_1	c_2	c_3	c_4	c_5	c_6	c_7
0.1	23.1	34.5	0.85	0.51	0.84	1.11	0.35
0.2	11.9	17.4	0.43	0.26	0.44	0.56	0.19
0.3	12.1	12.8	0.39	0.25	0.38	0.35	0

REFERENCES

1. Aumann R.J.,Integrals of set-valued functions.J.Math.Anal.Appl. Vol.12,1965,p.1-12.

2. Bank B.,Guddat J.,Klatte D.,Kummer B.,Tammer K.,Nichtlineare parametrische optimierung. Seminarbericht 31,Humboldt Universität Berlin,1981.

3. Berge C.,Espace topologique et fonctions multivoques.Dunod,Paris, 1965.

4.Binding P.Singularly perturbed optimal control systems.Part I: Convergence.SIAM J.Contr.Optim.,Vol.14,No.4,1976.

5. Bode H.W.,Network analysis and feedback amplifier design.D.Van Nostrand Co.,Princeton 1945.

6. Brunovský P.,Regular synthesis for the linear-quadratic optimal control problem with linear control constraints.J.Diff.Equations, Vol.38,No.3,1980.

7. Budak B.M.Vasilev F.P.,Some computational aspects of optimal control problems.Moscow Univ.Press,1975 (Russian).

8. Ciarlet P.G.,The finite element method for elliptic problems. North Holland,Amsterdam,New York,Oxford,1978.

9. Clarke F. A new approach to Lagrange multipliers.Math.of Oper. Research,Vol.1,No.2,1976.

10.Culum J.,Finite-difference approximations to state constrained optimal control problems.SIAM J.Control,Vol.10,1972,p.649-670.

11.Dmitriev M.G.,On the continuity of the solution of Mayer's problem with respect to singular perturbations.J.Vych.Math.and Math.Phys. Vol.12,No.3,1972 (Russian).

12.Dontchev A.L.,Sensitivity analysis of linear infinite-dimensional optimal control systems under changes of system order.Control & Cybern.,Vol.3,No.3-4,1974.

13. ———— ,A dual Ritz method for solving optimal control problems with equality state constraints.Archiwum Aut.i Telemech. Vol.23,1978,p.37-44.

14. ———— ,Error estimates for a discrete approximation to constrained control problems.SIAM J.Numer.Anal.,Vol.18,No.3,1981.

15. ———— ,Efficient estimates of the solutions of perturbed control problems.JOTA,Vil.35,No.1,1981.

16. ———— ,On the sensitivity of optimal control problems with phase constraints.Izvestija AN BSSR,Ser,Phys.-Math.,Vol.5,1981,p. 24-28 (Russian).

17. ———— ,On the order reduction of optimal control systems. Banach Centre Publications (to appear).

18. ———— ,Gičev T.R.,Convex singularly perturbed optimal control problem with fixed final state.Controllability and convergence.Math.Operationsforsch.und Stat.,Ser.Optimization,Vol,10,No.3, 1979.

19. ——— ,Veliov V.M., Singular perturbation in Mayer's problem for linear systems.SIAM J.Contr.Optim. (to appear).

20. ——— , ——— ,A singularly perturbed optimal control problem with fixed final state and constrained control.Control & Cybern. (to appear).

21.Dragan V.,Halanay A.,Suboptimal linear controller by singular perturbation technique.Rev.Roum.Sci.Techn.,Ser.Electr.Energ.,Vol.24, No.4,1976.

22. Eckeland I.,Temam R.,Convex analysis and variational problems. North Holland and American Elsevier,New York,1976.

23. Ermol'ev J.M.,Gulenko V.P.,Tzarenko T.I.,Finite element methods in optimal control problems.Naukova Dumka,Kiev,1978 (Russian).

24. Gabasov R.F.,Kirillova F.M., Qualitative theory of optimal processes.Nauka,Moscow 1971 (Russian).

25. Gamkrelidze R.V.,On the sliding optimal regimes.Soviet.Math. Dokl.,Vol.143,No.6,1962 (Russian).

26. Gičev T.R.,Well-posedness of optimal control problems with an integral convex performance index.Serdica,Vol.2,1976,p.334-342 (Russian).

27. ——— ,Singular perturbation in an optimal control problem with convex integral index - conditionally stable case.Serdica,Vol.7, 1981,p.306-319 (Russian).

28. ——— ,Dontchev A.L.,Linear optimal control system with singular perturbation and convex performance index.Serdica,Vol.4, 1978,p.24-35.

29. ——— , ——— , Convergence of the solutions of convex optimal control problems with time delay.Rendiconti di Matematica, Vol.11,No.6,1978.

30. ——— , ——— , Convergence of the solutions of the singularly perturbed time-optimal control problem.Prikl.Math.& Mech. Vol.43,1979,p.466-474 (Russian).

31.Girsanov I.V.,Lectures on the mathematical theory of extremum problems.Moscow Univ.Press,1970 (Russian).

32. Glizer V.J.,Dmitriev M.G.,On the convergence of the solutions of the analytic regulator design problem with singular perturbations. Prikl.Math.& Mech., Vol.41,No.3,1977 (Russian).

33. Gollan B.,Perturbation theory for abstract optimization problems. JOTA ,Vol.35,No.3,1981.

34. Hadamard J.,Lectures on Cauchy problem, Yale Univ.Press,1923.

35. Hager W.W.,The Ritz-Trefftz method for state and control constrained optimal control problems.SIAM J.Numer.Anal.,Vol.12,1975, p.854-867.

36. ——— ,Convex control and dual approximations.Control & Cybern.,Vol.8,1979,Part I,p.5-12,Part II,p.321-338.

37. ——— ,Rate of convergence for discrete approximations to unconstrained control problems.SIAM J.Numer.Anal.,Vol.13,1976,p.449-471.

38. ——— ,Lipschitz continuity for constrained processes.SIAM J.Contr.Optim.,Vol.14,1979,p.321-338.

39. ——— ,Mitter S.K.,Lagrange duality theory for convex control problems.SIAM J.Contr.Optim.Vol.14,1976,p.883-856.

40. Huard P.(Ed.).Point-to-set maps and mathematical programming. Mathematical Programming Study 10,1979.

41.Joffe A.D.,Generalized solution of systems with control.Diff. Uravn.,Vol.5,No.6,1969 (Russian).

42. Joffe A.D.,Tikhomirov V.M.,Theory of extremum problems.Nauka, Moscow,1974 (Russian).

43.Kirillova F.M.,On the continuous dependence of the solutions of an optimal control problem with respect to initial data and parameters.Uspekhi Math.Nauk,Vol.17,No.4 (106),1962 (Russian).

44.Kokotovič P.V.,O'Malley R.E.,Jr.,Sannuti P.,Singular perturbation and order reduction in control theory.Automatica ,Vol.12,No.2,1976.

45. ———— ,Alemong J.J.,Winkelman J.R.,Chow J.H.,Singular perturbation and iterative separation of time scales.Automatica, Vol.16,1980,p.23-33.

46. Lempio F.,Maurer H.,Differentiable stability of infinite-dimensional nonlinear programming.Appl.Math.Optim.,Vol.6,No.2,1980.

47. Levikov A.A.,On limit properties of dynamic systems with convex constraints.J.Vych.Math.and Math.Phys.,Vol.19,No.4,1979 (Russian).

48. Levitin E.S.,On the local perturbation theory of mathematical programming in a Banach space.Soviet.Math,Dokl.Vol.16,No.5,1975.

49. Lions J.L.,Controle optimal de systemes gouvernes par des equations aux derivees partilles. Dunod,Gauther-Villars,Paris,1968.

50. Luenberger D.G.,Optimization by vector space method. J.Wiley, New York,1969.

51. Malanowski K.,On convergence of finite-difference approximations to optimal control problems for systems with control appearing linearly.Archiwum Aut.i Telemech.,Vol.24,1979,p.155-171.

52. ———— ,On convergence of finite-difference approximations to control and state constrained convex optimal control problems. Archiwum Aut.i Telemech.,Vol.24,1979,p.319-337.

53. ———— ,Convergence of approximations vs. regularity of solutions for convex control-constrained optimal control problems. Appl.Math.Optim.,Vol.8,1981,p.69-95.

54. Moiseev N.N.,Mathematical problems of the system analysis. Nauka, Moscow,1981 (Russian).

55. Mordukhovič B.Sh.,Existence of optimal control.In Sovremennye Probl.Math.,ed.R.V.Gamkrelidze,Vol.6,1976 (Russian).

56. ———— ,On the finite-difference approximation to optimal control systems.Prikl.Math.and Mech.,Vol.42,1978,p.431-440 (Russian).

57. O'Malley R.E.,Jr.,Boundary layer method for certain nonlinear singularly perturbed optimal control problems.J.Math.Anal.Appl. Vol.45,1974,p.468-484.

58.Optimization od a gass transmission system in nonstationary conditions.Technical report.Institute of Automatic Control,Techn,Univ. of Warsaw,1979 (Polish).

59. Peng T.K.C.,Invariance and stability for bounded uncertain systems.SIAM J.Control,Vol.10,No.4,1972.

60. Petrov N.N.,On the well-posedness of the time-optimal control problem.Diff.Uravn.,Vol.13,No.11,1977 (Russian).

61. Poljak B.T.,Existence theorems and convergence of minimizing sequences for constrained extremum problems.Soviet Math.Dokl.,Vol.166, No.2,1966 (Russian).

62. Pontrjagin L.S.,Boltianskij V.G.,Gamkrelidze R.V.,Mishchenko E.F., Mathematical theory of optimal processes.Nauka,Moscow 1969 (Russian).

63. Reddien G.W.,Collocation at Gaus points as a discretization in optimal control.SIAM J.Contr.Optim.,Vol.17,No.2,1979.

64. Robinson S.M.,Stability theory for systems of inequalities.SIAM J.Numer.Anal.,Part I:Linear systems,Vol.12,No.5,1975;Part II:Differential nonlinear systems,Vol.13,No.4,1976.

65. Rockafellar R.T.,State constraints in convex control problems of Bolza,SIAM J.Control,Vol.10,No.4,1972.

66. Rolewicz S.,Funktionalanalysis und stuerungstheorie.Springer-Verlag,Berlin,Heidelberg,New York,1976.

67. Rosenwasser E.N.,Yusupov R.M.,Sensitivity of control systems. Nauka,Moscow 1981 (Russian).

68. Sendov Bl.,On a problem of G.I.Marchuk,in Numerical methods of math.phys.,ed.G.I.Marchuk,Novosibirsk,1979,p.5-10 (Russian).

69. Tchukanov S.V.,On the continuous dependence of the solutions of an economic planning problem of problems conditions.Autom.i Telemech.,Vol.5,1978,p.113-121(Russian).

70. Tichonov A.N.,Systems of differential equations containing a small parameter in the derivative.Math.Sbornik, Vol.31(73),No.3,1952 (Russian).

71. Vasilev F.P.,Lecture notes on methods for solving extremum problems.Moscow Univ.Press,1972 (Russian).

72. Vasileva A.B.,Butuzov V.F.,Asymptotic expansions of solutions of singularly perturbed equations. Nauka,Moscow 1972 (Russian).

73. ——— ,Dmitriev M.G.,Singular perturbations in optimal control problems,in Itogi nauki i techn.,Vol.20,1982,p.3-78 (Russian).

74. Vladimirov A.B.,Nesterov J.E.,Chekanov J.N.,On uniformly convex functionals.Vestnik Mosk.Univ.,Ser.Vych.Math.Kibern.,Vol.2,1978, p.12-23 (Russian).

75. Warga J.,Relaxed variational problems.J.Math.Anal.Appl.Vol.4, No.1,1962.

76. ——— ,Optimal control of differential and functional equations.Academic Press,New York 1976.

77. Wierzbicki A.P.,Unified approach to the sensitivity analysis of optimal control systems.Proc.IV Congr. IFAC, Warszawa 1969.

78. ——— ,Models and sensitivity of control systems.WNT Warszawa 1977 (Polish).

79. ——— ,Dontchev A.L.,Basic relations in the performance sensitivity analysis of optimal control systems.Control & Cybern., Vol.4,No.3-4,1974.

80. Wonham W.M.,Linear multivariable control: a geometric approach. Nauka,Moscow 1980 (Russ.transl.).

81. Zolezzi T.,A characterization of well-posed optimal control systems.SIAM J.Contr.Optim.,Vol.19,No.5,1981.

82. ——— ,Approximations and perturbations of minimum problems. (to appear).

INDEX

Lecture Notes in Control and Information Sciences

Edited by A. V. Balakrishnan and M. Thoma

Vol. 22: Optimization Techniques
Proceedings of the 9th IFIP Conference on Optimization Techniques,
Warsaw, September 4–8, 1979
Part 1
Edited by K. Iracki, K. Malanowski, S. Walukiewicz
XVI, 569 pages. 1980

Vol. 23: Optimization Techniques
Proceedings of the 9th IFIP Conference on Optimization Techniques,
Warsaw, September 4-8, 1979
Part 2
Edited by K. Iracki, K. Malanowski, S. Walukiewicz
XV, 621 pages. 1980

Vol. 24: Methods and Applications in Adaptive Control
Proceedings of an International Symposium
Bochum, 1980
Edited by H. Unbehauen
VI, 309 pages. 1980

Vol. 25: Stochastic Differential Systems – Filtering and Control
Proceedings of the IFIP-WG7/1 Working Conference
Vilnius, Lithuania, USSR, Aug. 28 – Sept. 2, 1978
Edited by B. Grigelionis
X, 362 pages. 1980

Vol. 26: D. L. Iglehart, G. S. Shedler
Regenerative Simulation of Response Times in Networks of Queues
XII, 204 pages. 1980

Vol. 27: D. H. Jacobson, D. H. Martin, M. Pachter, T. Geveci
Extensions of Linear-Quadratic Control Theory
XI, 288 pages. 1980

Vol. 28: Analysis and Optimization of Systems
Proceedings of the Fourth International Conference on Analysis and Optimization of Systems
Versailles, December 16–19, 1980
Edited by A. Bensoussan and J. L. Lions
XIV, 999 pages. 1980

Vol. 29: M. Vidyasagar,
Input-Output Analysis of Large-Scale Interconnected Systems –
Decomposition, Well-Posedness and Stability
VI, 221 pages. 1981

Vol. 30: Optimization and Optimal Control
Proceedings of a Conference Held at Oberwolfach, March 16–22, 1980
Edited by A. Auslender, W. Oettli, and J. Stoer
VIII, 254 pages. 1981

Vol. 31: Berc Rustem
Projection Methods in Constrained Optimisation and Applications to Optimal Policy Decisions
XV, 315 pages. 1981

Vol. 32: Tsuyoshi Matsuo,
Realization Theory of Continuous-Time Dynamical Systems
VI, 329 pages, 1981

Vol. 33: Peter Dransfield
Hydraulic Control Systems –
Design and Analysis of Their Dynamics
VII, 227 pages, 1981

Vol. 34: H.W. Knobloch
Higher Order Necessary Conditions in Optimal Control Theory
V, 173 pages, 1981

Vol. 35: Global Modelling
Proceedings of the IFIP-WG 7/1 Working Conference Dubrovnik, Yugoslavia,
Sept. 1–5, 1980
Edited by S. Krčevinac
VIII, 232 pages, 1981

Vol. 36: Stochastic Differential Systems
Proceedings of the 3rd IFIP-WG 7/1 Working Conference
Visegrád, Hungary, Sept. 15–20, 1980
Edited by M. Arató, D. Vermes, A.V. Balakrishnan
VI, 238 pages, 1981

Vol. 37: Rüdiger Schmidt
Advances in Nonlinear Parameter Optimization
VI, 159 pages, 1982

Vol. 38: System Modeling and Optimization
Proceedings of the 10th IFIP Conference
New York City, USA, Aug. 31 – Sept. 4, 1981
Edited by R.F. Drenick and F. Kozin
XI, 894 pages. 1982

Vol. 39: Feedback Control of Linear and Nonlinear Systems
Proceedings of the Joint Workshop on Feedback and Synthesis of Linear and Nonlinear Systems
Bielefeld/Rom
XIII, 284 pages. 1982

Vol. 40: Y.S. Hung, A.G.J. MacFarlane
Multivariable Feedback:
A Quasi-Classical Approach
X, 182 pages. 1982

Vol. 41: M. Gössel
Nonlinear Time-Discrete Systems –
A General Approach by Nonlinear Superposition
VIII, 112 pages. 1982

Vol. 42: Advances in Filtering and Optimal Stochastic Control
Proceedings of the IFIP-WG 7/1 Working Conference
Cocoyoc, Mexico, February 1–6, 1982
VIII, 391 pages. 1982

Lecture Notes in Control and Information Sciences

Edited by A. V. Balakrishnan and M. Thoma

Vol. 43: Stochastic Differential Systems
Proceedings of the 2nd Bad Honnef Conference
of the SFB 72 of the DFG at the University of Bonn
June 28 - July 2, 1982
Edited by M. Kohlmann and N. Christopeit
XII, 377 pages. 1982.

Vol. 44: Analysis and Optimization of Systems
Proceedings of the Fifth International
Conference on Analysis and Optimization of Systems
Versailles. December 14-17, 1982
Edited by A. Bensoussan and J. L. Lions
XV, 987 pages, 1982

Vol. 45: M. Arató
Linear Stochastic Systems
with Constant Coefficients
A Statistical Approach
IX, 309 pages. 1982

Vol. 46: Time-Scale Modeling of Dynamic Networks
with Applications to Power Systems
Edited by J. H. Chow
X, 218 pages. 1982

Vol. 47: P.A. Ioannou, P.V. Kokotovic
Adaptive Systems with Reduced Models
V, 162 pages. 1983

Vol. 48: Yaakov Yavin
Feedback Strategies for Partially
Observable Stochastic Systems
VI, 233 pages, 1983

Vol. 49: Theory and Application of Random Fields
Proceedings of the IFIP-WG 7/1
Working Conference
held under the joint auspices of the
Indian Statistical Institute
Bangalore, India, January 1982
Edited by G. Kallianpur
VI. 290 pages. 1983

Vol. 50: M. Papageorgiou
Applications of Automatic Control Concepts
to Traffic Flow Modeling and Control
IX, 186 pages. 1983

Vol. 51: Z. Nahorski, H.F. Ravn, R.V.V. Vidal
Optimization of Discrete Time Systems
The Upper Boundary Approach
V, 137 pages 1983

Vol. 52: A. L. Dontchev
Perturbations, Approximations and Sensitivity Analysis
of Optimal Control Systems
IV, 158 pages. 1983